Never to Return Home

John M. Tatum and the Fourth Texas Infantry Regiment

ISBN: 978-1-60800-005-0
Library of Congress Control Number: 2011902725

Design and layout by Mike Lovell
American Historical Publishing
Lancaster, PA

717 239 0490
www.americanhistoricalpublishing.com

Printed by Yurchak Printing Inc.
Lancaster, PA

www.yurchak.com

Designed, published and printed in the
United States of America

Never to Return Home

John M. Tatum and the Fourth Texas Infantry Regiment

Timothy C. Tatum

Chaplain (Colonel) US Army Retired

Preserving Our Heritage
One Story at a Time

American Historical Publishing
Lancaster, PA

TEXAS

Dedication

To my many aunts and uncles who would have enjoyed this book about their grandfather and his regiment.

I wish that I had been given the foresight to write this history during their time on earth.

General Robert E. Lee, on the back porch of his Richmond home, just one week after surrendering his army at Appomattox.
Photographed by Matthew Brady

Contents

Cover Image: The painting on the cover depicts the charge made by General Hood and the Fourth Texas Infantry Regiment at Gaines Mill. The painting was commissioned by the author and his brother Tom in honor of their great-grandfather. The author worked closely with the artist, Dale Gallon, on the details to ensure total accuracy. The soldier with the rifle to the left of General Hood, the viewers right, is John Tatum. Although no pictures exist of John Tatum, the likeness was taken from a picture of John's son.

Limited editions of the print are available from www.gallon.com.

Acknowledgments and Observations

A casual visit to Military History Institute in Carlisle, Pennsylvania, spawned the idea for this book. I was a student at the US Army War College from 1985 to 1986. During the first few months of my student year I had not ventured across the street from the college to visit the Institute. Embarrassed by my lack of interest, I decided to comb through their Civil War section. I was armed with very little information about my great-grandfather, John Tatum, but it was a starting point. I had been told by my uncles that he died during the Civil War fighting for a Texas unit but they knew nothing about the circumstances of his death, where he fought, how he died, or the unit to which he was assigned. To my amazement, it was less than one hour when I discovered his unit, the battles he endured, and when he died. A later visit to the National Archives during my three years on the War College faculty revealed John's military record. The record clearly recorded the approximate date John joined the 4th Texas Infantry Regiment, the date of his death, and the cause of death. My brother Tom and I decided that we should do something with this information by commissioning a friend and famous artist, Dale Gallon, to paint a picture of John Tatum making the charge at Gaines Mill with General Hood. Tom financed the painting and the limited edition prints while I researched the details of the attack for the artist.

I knew that the information needed to be expanded and turned into a book. The Army interrupted my plans by sending me to a new assignment as the senior chaplain for the US Army Pacific Command. It wasn't until several years after my retirement from the military that I determined to finish this project. Fortunately the creation of the internet, with the addition of numerous census records, aided me greatly in my task. My greatest regret was not completing this project while my aunts and uncles were still living. They would have enjoyed learning about their grandfather's exploits during the most challenging time in our nation's history.

I have several people to thank for their help in getting this book into its final form. My wife, Carolyn, was a constant source of encouragement and made numerous helpful suggestions. Several others assisted by proofing the first and second drafts. They include my son, Mark, and my sister-in-law, Sherry Bradshaw. Finally, a special thanks to Jack Tatum for supplying the pictures seen in Appendix B.

A fairly complete Tatum family tree can be found on Ancestry.com thanks to the hard work of Diane Tatum. A special thanks to Diane for the numerous hours that she spent researching family connections and dates and entering data into a software program (Family Tree Maker).

Copies of the database which produces the tree are available from the author. A copy of Family Tree Maker is required to see and print the Tatum family tree.

As the author, I take full responsibility for the contents and apologize in advance for any mistakes in names, dates or places. Suggestions for and improving the tree and any corrections or additions can be emailed to the author at tim_tatum@comcast.net.

I hope the readers of the following pages have a chance to walk some of the battlefields where John and his companions paid such a high price. You will find a suggested itinerary for visiting the battlefields at Appendix C.

Tim Tatum
2011

Introduction

We are still recovering from the profound impacts made by the Civil War, even after 145 years. In the world of religion, denominations remain split that were once whole and moving in the same direction. Families that went their separate ways have lost contact with one another. There are still households as far north as Maryland that still fly the confederate flag in their front yard. Some of the scars will probably remain for another hundred years.

The tragedies that came out of this war profoundly changed the Tatum family. The death of John M. Tatum in Richmond, and the death of his wife Margaret not too many months prior, sent his children in several new directions. James grew up not knowing his older sister, let alone his stepbrother and sister. To the best of our knowledge, the children never came together again once they were farmed out to various families.

This story is told about the Civil War exploits of John Tatum and his company, B Company, 4th Texas Infantry Regiment. Fortunately there are a number of first-hand, or primary, resources for telling the story. The bibliography will outline all of the resources consulted but the books and papers that were the most critical to this story include: *Rags and Hope,* by Val Giles, a member of B Company, 4th Texas Infantry; *Hood's Texas Brigade* by J. B. Polley, a member of F Company, 4th Texas Infantry; *Chaplain Davis and Hood's Texas Brigade* by Chaplain Nicholas A. Davis, chaplain for the 4th Texas Infantry; *Advance and Retreat* by General John Bell Hood; and several books by Harold Simpson, a retired military colonel and a former acquaintance of the author.

Over 623,000 Americans died in the Civil War. Not even World War II approaches this figure: between 1941 and 1945 US forces sustained 407,000 fatal casualties. Total casualties in the Mexican War had been just over 13,000, and fewer than 2000 of these were battle deaths. The US lost fewer than 5000 men during the entire American Revolution. While 65 to 70 percent of the Civil War deaths were caused by disease (such was the case for John Tatum), the carnage inflicted by mid-nineteenth century weapons was devastating. Both sides began the war with volunteer armies. Most of the leadership predicted that a decisive result would come by the summer of 1861. Battle followed battle and Lee's victories bought the south time, but they could not inflict a fatal blow on the northern army.

Civil War battles were generally dominated by infantry armed with rifled muskets. The Springfield .58-caliber and similar weapons inflicted the majority of battlefield casualties. Many soldiers, including some Texans, fought with smooth-bores that were very limited in accuracy and range. Whether smoothbores, or rifled muskets, firepower had to be concentrated in order to be effective. Manuals that were originally developed by the French (often referred to as Napoleon tactics) were used by both sides. The soldiers were instructed to form two lines, more or less shoulder to shoulder and form on their flag. This line of battle was usually preceded by skirmishers who would "feel" for the enemy line. Only a dense line of troops could develop firepower to form a strong defense and the momentum to break through in the offense.

The Civil War infantry regiments varied greatly in size and composition. Generally the regiment was divided into 10 companies, each 40 to 100 strong. On paper a regiment could take 900 to 1000 men into battle. In reality, few regiments could muster more than 450 men for action. The companies were designated with a letter, with A Company in the place of honor or the right flank. B Company was normally assigned the left flank. In the center stood the regimental colors, which served as a rallying point in the defense and proper alignment of the lines in the offense. Bearing the colors was the riskiest job in the regiments. Their job was almost hopeless but the Texans never lacked for immediate volunteers.

Infantry companies were commanded by a captain, a 1st lieutenant, a 2nd lieutenant, a 1st sergeant, four sergeants, and eight corporals. Most of the time the Texas companies were short of officers, sergeants, and privates. Companies included two musicians, typically a trumpet player or bugler and a drummer. Musicians from the various companies formed the regimental, or in some cases, the brigade band. The two-rank formation was difficult to maintain, turn, and maneuver in the heat of battle. The soldiers spent far more time practicing their drill movements than they did firing their rifles. If a company or regiments could not maintain its cohesion, it was doomed to defeat or heavy losses.

Weapons

As mentioned earlier, the rifled muskets were far more effective weapons than the smoothbores they replaced. In the hands of a Texan they were capable of deadly accuracy at several hundred yards. Accurate fire was almost impossible in the heat of battle when standing in the skirmish line. A regiment might start a fight with a well-timed volley but it was usually followed by firing at will. Soldiers carried 60 to 100 rounds into battle.[1] The fouling from the black powder accumulated rapidly in the barrel. It only took 20 to 30 shots to choke the barrel so badly that the rammer had to be hammered down to get the ball fully seated. It only took 40 rounds to produce a bruised and battered shoulder.

The intensity and noise of battle made for many mistakes in the use of these weapons. Some 37,000 muskets were salvaged from the

---

1. Cartridges were made of black powder wrapped in thick paper folded at both ends. The soldier tore the end of the cartridge with his teeth, poured it in the barrel, added a lead ball, rammed the contents to the bottom of the barrel, placed a cap under the hammer, and fired. A good soldier could load and fire three rounds in a minute.

Gettysburg battlefield. Of these, 24,000 were loaded, three-fourths of them with more than one round. The 6000 with just one round included many weapons with the whole cartridge rammed down unopened, or with the bullet behind the powder charge instead of in front.[2]

Tactics

Officers knew that it was critical that the first and second volleys be delivered at the right time and the right place, preferably close to the enemy. The outcome of the battle depended on getting this right. In order to ensure the desired results, officers led their companies and regiments into battle. Soldiers faced the risk of shooting their officers in the back if they fired early and without command. A good example of this tactic was demonstrated by General Hood when he led the 4th Texas Infantry into battle at Gaines Mill.

Cavalry units were used as a screening force for the purpose of determining the enemy's location, size, and movement. They were the eyes and ears of the division and corps commander. J. E. B. Stuart was assigned this role in support of General Lee and the Army of Northern Virginia. Military officers still debate as to how the battle of Gettysburg would have been different if J. E. B. Stuart had not been late.[3] At the regimental level and at much shorter distances, infantry units protected themselves by sending skirmishers ahead of the main body. Two of the ten companies would be sent ahead but spread in a much wider formation so they covered the whole front. A 450 man regiment would occupy 150 yards if all 10 companies were in the two-deep formation. If two companies went forward as skirmishers, the line would shrink to 120 yards.[4]

---

2. Figures taken from *The Civil War Military Machine* by Ian Drury and Tony Gibbons, page 24.
3. The Confederates generally had the better horsemen and cavalry and dominated the cavalry battles for the first two years of the Civil War.
4. The frontage of a regiment formed in line of battle was calculated at two feet per man

As the Civil War progressed into its final years, trench warfare became the common tactic. The increasing accuracy of the rifled muskets, introduction of repeating rifles, devastating artillery rounds, and better trained solders demanded a safer way of fighting. By the end of the war, the Confederates dug extensive trenches near Richmond and Petersburg, Virginia.[5] Prior to this, Federals and Confederates in defensive positions felled trees and dug rifle pits to reinforce their positions. Whenever possible the defensive position was constructed on the side of a hill or ridge with supporting artillery occupying the higher ground.

Artillery

The effective use of artillery is one of the main reasons for entrenching. The field batteries were organized into four and six gun batteries. Generally the Union used six gun batteries, while the Confederates varied between four and six. Batteries were attached to infantry brigades, divisions, and corps, so each infantry formation had its own artillery.

The use of rifled cannon increased the effective range of artillery in the same manner as the rifled musket. This extended range was demonstrated at Gaines Mill when the Union artillery engaged the Confederate infantry at 2½ miles. The Confederates had several different cannons within their batteries. It was normal for a four-gun battery to consist of a pair of 12-pounder Napoleon smoothbores and a pair of 10-pounder Parrott rifles. Needless to say, the variety of artillery pieces made for a logistic nightmare. The artillery's most effective projectile was the case-shot made out of tin cylinders or linen bags packed with musket balls. Its effective range was about 300 yards, producing a shotgun effect. Case-shot could blow large holes in an advancing line of infantry.

The best marksmen in an advancing infantry unit would attempt to disable the artillerymen attending the nearest battery. Even the horses used to draw the batteries were considered a worthy target.

5. To a great extent, this new tactic dictated the way World War I would be fought.

Capturing an artillery piece, or an enemy regimental flag, was considered a major accomplishment and worthy of a major prize. Some in the Union Army were awarded a Congressional Medal of Honor for dragging an enemy flag or artillery piece off the field.

Editor's notes:

1. The references quoted throughout the book include a number that were written during the period of the Civil War. In these cases the author has preserved the original spelling, which sometimes differs from today's usage. E. g. spelling "halted" as "haulted;" "Indianapolis" as "Indianapoilis."
2. In another usage of language of the period we see "...the easy going Lieutenant Hughes of the 4th who did not have much practice at cussing, yelled, 'Go it, Stringfield—go it! Kill him, dod damn it, kill him."

   People of the period were more conservative than is common today. It was not uncommon for a speaker to use a surrogate word such as "dod" for "god;" similarly, in this instance it may have been the writer who made this recollection of Lt. Hughes exhortation.
3. One of the principal sources for this book is *Chaplain Davis and Hood's Texas Brigade* Chaplain Nicholas A. Davis, (San Antonio: Principia Press of Trinity University, 1962). This book differs from the other sources of the period in that it was written *during* the war, whereas most others were written after the event. Consequently, Chaplain Davis's book conveys a more optimistic view about the outlook of the war.

PROCLAMATION

BY THE GOVERNOR OF

THE STATE OF TEXAS

WHEREAS, by the several acts of the Legislature of the State of Texas, hereto appended, it is declared that an Ordinance dissolving the connection of Texas with the Federal Union, shall be submitted to the people of Texas, for their ratification or rejection; and whereas, it is also declared; that an election shall be held on the 23d day of February, A. D. 1861, at which the qualified electors of the State shall vote upon the same, and it is made the duty of the Governor to issue Proclamation for the same.

NOW THEREFORE, I, SAM HOUSTON, Governor of the State of Texas, in obedience to law, do issue this my Proclamation, directing the Chief Justices of the various counties of the State, (or in case of vacancy in the office of Chief Justice, or the failure or inability of that officer to act, then any two of the County Commissioners) to give due notice and cause polls to be opened at the several precincts of their respective counties on the said 23d day of February, A. D., 1861, at which the qualified electors will signify their approval or disapproval of said Ordinance, by the use of the words, "FOR SECESSION" or "AGAINST SECESSION"

And the Chief Justices or County Commissioners, as the case may be, shall cause returns to be made to them of said election, on or before the 26th day of February, A. D. 1861, and they shall on that day make duplicate returns of the same, as provided for in the two acts of the Legislature hereto appended.

{ L. S. } Given under my hand and the Great seal of the State at Austin, this the 9th day of February, A. D. 1861, and of the Independence of the United States, the Eighty-fifth, and of Texas, the Twenty-fifth,

By the Governor, **SAM HOUSTON.**

E. W. CAVE, Secretary of State.

Chapter One

Texas Goes to War

Delegates to the Secession Convention of Texas met on February 1, 1861, to decide the future fate of the Lone Star state. Over the next four days they debated whether or not to secede from the Federal Union. In spite of governor Sam Houston's seemingly pro Unionist views, the assembly voted 166 to 8 to secede from the Union. The convention ended on February 4 and on February 23 put the matter to a vote of the people. The people of Texas ratified the vote by a three-to-one majority, 46,129 to 14,697.[1]

The results of the Proclamation were announced to the people on March 4 with much shouting and fanfare. Texas, at least for one day, became an independent nation. The Texas Convention approved an ordinance on March 5 tying Texas to the newly formed Confederate States of America. This action was superseded by the Confederate Provisional Congress in that they passed an act on March 1, 1861, to admit the Lone Star state.[2] Texas became the seventh state to join the southern cause.[3]

The first major problem facing this new Confederate state was what to do about the numerous federal troops garrisoned through-

1. Harold Simpson, *Hood's Texas Brigade: Lee's Grenadier Guard* (Dallas: Alcor Publishing Company, 1983), p. 2. (quoting from the original source: Winkler, *Secession Journal*, p. 49.)

2. . *Ibid.* Winkler, *Secession Journal*, pp. 101-102.

3. South Carolina was certain that fifteen states would join the Confederacy. The final number, fourteen, fell one state short of their estimate.

out the state. A large number of these soldiers were patrolling the Texas/Mexico border where they guarded against roving bandits and Indian raiding parties. Fortunately the senior military commander was sympathetic to the South. He managed to reposition the federal forces to other areas outside of the state without a shot being fired. This created a serious vacuum along the border that Texas tried to fill with its own volunteers. Border protection during the Civil War never reached the level sustained by the Federal soldiers. Problems in West and Southwest Texas became a nagging issue until the end of the war.

According to Harold Simpson:

> Several times during the early months of 1861, the Federal government contemplated military intervention in Texas. On at least two occasions early in the year, Abraham Lincoln offered to send Federal troops into the Lone Star state and place them under Sam Houston's command. . . . After consulting with four Texas Unionist leaders, Houston decided against it saying, "Gentlemen, I have asked your advice and will take it, but if I were twenty years younger I would accept Mr. Lincoln's proposition and endeavor to keep Texas in the Union."[4]

Texas eventually mobilized 50,000 to 60,000 for the war effort, many of these remaining in Texas to protect the home front. The most famous were the 4,000 that formed the thirty-two Texas companies that were organized into the 1st, 4th, and 5th Texas Infantry Regiments. All three regiments became the heart of Hood's Texas Brigade. Although this has been debated over the last 140 years, many outside of Texas have declared this brigade to be the best and fiercest group of soldiers on either side during the entire Civil War. General Lee declared on more than one occasion, "Where are my Texans? I need them."

Only a month after joining the Confederacy, the governor of Texas received two letters asking for 8,000 troops. It was the governor's responsibility to make sure these soldiers were equipped and drilled. A third letter arrived in May, 1861, asking for 2,000 additional infan-

4. Simpson, *op. cit.*, p. 7.

Val Giles, Company B, 4th Texas Infantry Regiment
Image from American Military Equipage, Vol. II

trymen. The companies that later became part of the 1st Texas Infantry Regiment were organized under the first levy calling for 8,000 troops. The 4th and 5th Infantry Regiments were organized as part of the 2,000 levy. The companies themselves were volunteers from the various towns and counties in East and Central Texas. The companies of 1st Texas were from East Texas, while companies of the 4th and 5th Texas were all from Central Texas.[5]

A few of the Texas companies were organized and partially trained prior to the war. Two examples were the "Bayou City Guards" which later became Company A, 5th Texas Infantry, and the "Tom Green Rifles," later known as Company B, 4th Texas Infantry. The Tom Green Rifles were organized at Austin in 1858, three years prior to the outbreak of the war. Company B became John Tatum's company but not before they moved to Virginia and faced their first encounter with the North. The Confederate government reserved the right to appoint the regimental officers once the companies arrived in Richmond.

Colonel Simpson quotes from a private letter in describing how the Tom Green Rifles (later B Company) obtained their uniforms:

5. Chaplain Nicholas A. Davis, *Chaplain Davis and Hood's Texas Brigade* (San Antonio: Principia Press of Trinity University, 1962), pp. 32-33.

> In some cases local citizens raised funds to help outfit the military companies. At Austin on May 20, "a group of Tennesseans met over the store of Darden and Maynard and contributed $1,000 to equip Captain B. F. Carter's company" –the "Tom Green Rifles" (company B, 4th Texas Infantry).[6]

Mary Lasswell, the editor of a book written by Val C. Giles, Company B, 4th Texas, described the uniforms worn by the "Tom Green Rifles:"

> They were dressed in four different shades of gray trimmed in black braid. The citizens of Austin had contributed money for the cloth; a local tailor had taken the measurements and cut out the uniforms, and the ladies' "Needle Battalion" had made them up. Headgear was left to the choice of the individual soldiers, and at least one recruit appeared with a stovepipe beaver."[7]

Governor Edward Clark established camps for training the various companies. Camp Clark served as the training site for several of the companies of the 4th Texas Infantry regiment. Camp Clark was located on the San Marcos River near the Hays-Guadalupe county line. General Tom Green, the namesake of the "Tom Green Rifles," served as the camp commandant. General Green became famous during the Texas Revolutionary War and Mexican wars.[8] Most of the other companies had very little training before leaving their home towns.

The various companies of the 1st Texas Infantry made their way to Virginia as best they could. There was no major plan that guided their movement. Because of their early departure, they were able to arrive in Richmond ahead of the other companies. The companies that later comprised the 4th and 5th Texas rendezvoused at Camp Van Dorn, a few miles east of Houston.

---

6. Simpson, *op. cit.*, p. 15.

7. Mary Lasswell, ed., *Rags and Hope: Memoirs of Val C. Giles, Four Years with Hood's Brigade, Fourth Texas Infantry, 1861-1865* (Harrisonburg, Virginia: Sprinkle Publications, 1887), pp. 23-24. See the photograph of Val Giles and his unusual hat.

8. *Ibid.*, p. 20.

Chaplain Davis gives this uninviting view of the camp:

> It was midsummer when the troops were taken to the camp, on Buffalo Bayou. The camp was in a low miasmatic, unhealthy region, and many of the men here contracted disease, from which they never recovered. So exhaustive was the climate and the place on the constitution of the men, that very little was done towards drilling them, and at the expiration of the time spent at that place, little or no improvement was discernible. The time spent there was spent most disagreeably, and many were the anathemas indulged in by the men at the cause of delay. While in this camp much kindness was shown the troops by the citizens of the neighborhood, and by those in the city of Houston.[9]

The first companies left Camp Van Dorn in smaller company segments, leaving their temporary camp a few days apart. The first section to leave consisted of five companies under the command of Captain J. C. G. Key. The group included Captain B. F. Carter's "Tom Green Rifles." Carter was a remarkable soldier even though he was trained as a lawyer. Later in the war he was promoted to Lieutenant Colonel and commander of the 4th Texas Infantry Regiment.

Chaplain Davis, writing in early 1863 and two years before the end of the Civil War, says of Captain Carter:

> B. F. Carter, 4th Texas, is from Maury County, Tennessee, and in his thirty-second year. He was graduated at Jackson College, Columbia, in 1850; and spent the years 1851-52 in the Law School at Cumberland University, Lebanon, when he procured a license and moved to the city of Austin in the fall, where he has since resided, and held various municipal offices, viz: Alderman, city Attorney, and Mayor two terms. . . . He commanded the 4th at Freeman's Ford, Manassas and Sharpsburg, and especially in the last two, his regiment found him to be everything that was necessary on those trying occasions. On both fields the struggle was long and bitter. Each one had witnessed the setting of the second sun before the enemy gave up the strife. And it is under circumstances like these that the best material is put to the severest test. But his gallantry and

9. Davis, *op. cit.*, p. 34.

> endurance was only equaled by the bravery and unalterable determination of the men under his command.[10]

Captain Key was an attorney from South Carolina who settled in Gonzales, Texas. He later commanded the 4th for a short time at Gaines Mill where he was severely wounded. He was wounded a second time at the battle of Gettysburg.[11]

The journey from Texas to Richmond could only be described as sheer misery. The men who survived the war never forgot what they had to go through as they passed through Louisiana. Chaplain Davis gives a vivid account of the long journey through the swamps and rivers, with little in the way of wagons. A portion of his account paints a vivid picture of what the men endured:

> Leaving the Calcasieu, our march was continued through a constant rain, through swamps and marshes, lagoons, wrecks, and every imaginable species of watery element, many of them over waist-deep, until we at length came to a halt, after a day's journey of 10 to 12 miles. We were halted in a prairie, immediately on a stream, whose waters were running out of its banks, and still rising. The wagons could not be crossed, and we went into camp to await the developments of the night. The next morning a rude bridge was constructed, over which the wagons were hauled by hand, while the animals swam across. The teams were then hitched up, and we proceeded eight miles to a stage stand, at a point called Pine Island, when we encamped late in the evening, the weather still continuing rainy.
>
> On the 24th we had the same sort of road, and similar weather for about 12 miles, to Welsh's Station. Here we crossed the stream on trees which we felled across it, and with some difficulty got our wagons over.

---

10. *ibid.,* p. 171. Unfortunately Carter died as the result of wounds he received while commanding the 4th Texas Infantry Regiment on July 2, 1863, in the Devil's Den while attacking "Little Round Top" at the battle of Gettysburg.

11. Val Giles recounts that "Of the ten original captains who went to Virginia with the 4th Texas Regiment in 1861, six of them were lawyers, two merchants, one a farmer and one a stockman. Of the thirty lieutenants, nearly one-third were lawyers. Of the fifty sergeants, fifteen were lawyers, and of the forty corporals about one-fifth were lawyers," p. 48.

> 25th. Came to the Mermenteau River. This day's march was, perhaps, the most severe on the trip. The distance traveled was not so great as on some other days, but at every step the toiling and wearied pedestrian encountered what appeared to be a little deeper and a little softer spot. It was on this day that we made the crossing of the "Grand Marias," or more aptly termed by the boys the "grand Miry." In many places the men waded up to the neck through the swamps, where the alligators lay basking in the tall grass, as if disputing the passage, and seemed reluctant to give way without a stern admonition in the way of a bayonet thrust, to impress them with a proper respect for the characters of the newcomers. Many were bayoneted by the soldiers, and held up in triumph as they went on plunging through the dangerous waters.[12]

The companies eventually arrived at a location where they could board box cars and ride the rails for the remainder of their journey. They arrived in Richmond on September 12 and were placed in a rough camp about three miles from the city. The remaining companies arrived over the next few weeks. Once everyone was in place, the soldiers began to speculate concerning the appointment of their regimental officers. According to Chaplain Davis, "The first attempt at giving a Colonel to the 4th Regiment was the appointment of R.T. P. Allen, a citizen of Bastrop, Texas, and the President of the Military Institute at that place."[13] His appointment was not acceptable to the men or the officers. The men simply felt he did not fit their view of a commander. Colonel Allen took the hint and returned to Texas.

The second attempt at making an appointment turned out to be more fortuitous. John Bell Hood was born in Owensville, Kentucky, June 29, 1831. He entered West Point in 1849 and graduated in 1853. He served for 22 months in the 4th Infantry in California. He transferred in 1855 to the 2nd Cavalry where then Lieutenant Colonel Robert E. Lee served as Hood's superior officer. In the winter of 1855-56 Hood served along the Texas frontier where he was

---

12. Davis, *op. cit.*, pp. 40-41.
13. *Ibid.*, p. 44.

wounded during a nasty engagement with the Indians. Impressed with its beauty and wide-open spaces, Hood adopted Texas as his home state. On the 16th of April, 1861, he resigned his commission under the United States Government and offered his services to the Confederacy.

Dyer says of John Hood:

> Under Lee he was a "blood young giant of fine military mien." At Gaines' Mill he had been superb, leading his Texas brigade in relentless charges that opened a gaping hole through which the rest of Lee's army rushed to victory—Lee's first important victory. This brought Hood's name to everyone's lips. And then there were Sharpsburg and Gettysburg and more fame as a brilliant and fearless combat officer. Between campaigns and battles he had been the toast of Richmond where hostesses considered their parties successful if the tall handsome cavalier attended. Choice food in the midst of increasing scarcity in Richmond was his, and charming women found him fascinating. After Chickamauga he had returned to Richmond to recuperate from the loss of his leg and now even more he was the idol of capital society. [14]

A number of the men knew Hood, or knew of him, during his time on the frontier. He was accepted by the enlisted men and officers without much opposition. John Marshall was appointed as the

---

14. John P. Dyer, *The Gallant Hood* (Indianapoilis: Bobbs-Merrill Co., 1950), p. 14. Hood lost an arm at Gettysburg. To be more accurate, the bone above his elbow was shattered and the division surgeon had no method for screwing the various pieces back together. Normally, the arm would have been removed but the surgeon decided to leave it in place where it hung at Hood's side the rest of his life. Later Hood had a leg amputated at Chickamauga. He finished the war by use of a special saddle that allowed him to be tied to saddle and horse. After the war, he married and took over General Longstreet's insurance business in New Orleans. He had numerous children but later died of Yellow Fever, along with his wife and one of his daughters. He became a Christian towards the end of the war in what has been called the Great Southern Revival during which thousands of southern soldiers and officers made decisions for Christ. For further reading on this subject see *The Great Revival in the Southern Armies* by William W. Bennett, and *Christ in the Camp* by J. William Jones. Both Bennett and Jones were Confederate chaplains.

Lieutenant Colonel. According to Val Giles he was the editor of the State Gazette, published at Austin, and a leading organ of the Democratic party. He was 50 years old when appointed to the regiment. He was born in Virginia and came to Texas in 1852. In 1858 he became the chairman of the Democratic State Convention. Because of his political connections, many of the men felt his appointment "was looked upon as savoring too much of political favoritism." When Hood made Brigadier General in March, 1862, Marshall became Colonel of the regiment. His service as regimental commander was cut short when he was killed while making the charge at Gaines Mill on June 27th, 1862.

The position of major was filled by Bradfute Warwick. His story is best told by a member of the 4th Texas.

> Bradfute Warwick received the appointment of Major. This appointment was regarded as a direct stab at our State pride. Warwick was altogether unknown to us, and was a Virginian, in no wise connected or identified with Texas or Texans. So his appointment was looked upon as an injustice to the men and to the State they came from.[15]
>
> A big kick was organized, but the older and cooler heads in the regiment prevailed over the younger men, so no steps were taken in opposition to his appointment. He was the son of Corban Warwick, and born in Richmond, Va., Nov. 24, 1839. And about twenty-three years of age when appointed Major of the 4th Texas.
>
> After attending the University of Virginia he entered the Medical College of New York, graduating from that institution at the age of nineteen. He then visited Europe and continued the study of medicine in Paris. Here his health began to fail and he was advised to travel. He visited Italy, Greece, and Turkey, Asia, and Africa and on his return through Italy he found that country in commotion. Fired by the spirit of freedom, he offered his services to Garibaldi, and was assigned to his medical staff. This position he soon resigned and enlisted as a private soldier, literally fighting his way up from a common soldier to a captaincy in six months. For his gallantry he

15. Lasswell, *op. cit.*, p. 43.

> received from the great dictator the Cross of the Legion of Honor.[16]
>
> Unfortunately, Warwick was one of the officers killed at Gaines Mill. The major made the charge near the head of company B, sword in hand, and a determination to see the Union on the run.

The officers and sergeants serving at the company level were selected prior to leaving Texas. As already noted, Captain B. F. Carter served initially as the commander of Company B; W. C. Walsh filled the position of 1st Lieutenant; J. T. McLaurin was appointed 2nd Lieutenant; and, R. J. Lambert, the man who would later recruit John Tatum, served as 3rd Lieutenant. The 5th Texas Infantry went through turmoil trying to agree on their regimental officers. Several in a row were run off by the men. The regiment finally agreed to the appointment of Colonel J. J. Archer as regimental commander, J. B. Robertson as Lieutenant Colonel, and P. J. Quattlebaum as major.

The 4th and 5th Texas continued to drill their ranks in proper battlefield tactics. The officers found it difficult to mold the independent spirit of the Texas men to the discipline of military life and orders. Their temporary camp, Camp Bragg, was the scene of daily close-order drills. According to Simpson:

> At Camp Bragg two things which were to play a vital but vastly different role in troop morale and welfare first made their appearance among the men of the 4th and 5th Texas Infantry—letters from home and body lice. As to their desirability, unfortunately, they appeared in reverse order. Letters from home came only spasmodically while the "graybacks," the popular name for lice, appeared in profusion. Mail from home, which was never received regularly even during the early part of the war, practically ceased to come when the Mississippi river fell into Federal hands in July 1863. But with the lice it was an entirely different story. They appeared in great numbers, and despite any temporary relief through scratching, picking, squeezing, and mashing, they were hard to control, and impossible to exterminate. This persistent little pest that was to annoy the Civil War armies would plague the Texans through-

16. *Ibid.*, pp. 45-46.

1st Texas Infantry in winter camp near Dumfries, Virginia
Photo courtesy US Army Military History Institute

out the war. To this miserable parasite the following prayerful ditty was dedicated by a wag in the Southern army:

Now I lay me down to sleep,
The Graybacks o'er by body creep;
If they should bite before I wake,
I pray the Lord their jaws to break.[17]

The 4th and 5th Texas set out to join the 1st Texas Infantry in late October, 1861, along the defensive lines just south of the Potomac River and near the small town of Dumfries. Due to several delays, they didn't set up camp until November 17. The Eighteenth Georgia Infantry Regiment commanded by Colonel Wofford joined the 1st, 4th and 5th Texas in performing defensive duties. The four regiments were formed into a Brigade known initially as Wigfall's Brigade but later became better known as Hood's Texas Brigade. Wigfall was a hard drinking lawyer and politician that migrated to Texas in 1848 where he later represented Texas in the United States Senate.

17. Simpson, *op. cit.*, p. 67.

Chaplain Davis describes the scene:

> As soon as it was understood we were quartered for the winter, the men set about building cabins; and it would puzzle any artist in the world to give the style of architecture in the cantonment, consequently we will not attempt the task. For some of them were on the hill, and some under the hill; some were on top of the ground and others were under the ground. Some were large, while others were small. One was in this shape, and another in that shape.[18]

No major engagements took place during the winter months of 1861-62. Several stories are told of Colonel Wigfall sounding the alarm at very inconvenient times when he thought the Federals were coming in force. The other officers soon learned that the colonel was easily alarmed and rarely had his facts correct. Small scouting parties regularly crossed the river in both directions, mostly to break the boredom. The weather throughout the winter was miserable and the roads were muddy. The officers gave up on regular drills while the soldiers concentrated on surviving the winter.

Based on the reports from the soldiers, officers, and Colonel Hood himself, "widespread disease, sickness, and epidemics swept the Confederate camps." During this time thousands were hospitalized, hundreds died, and hundreds more were so weakened that they had to spend months on convalescent leave or were discharged from the service. Typhoid fever, pneumonia, and dysentery, in particular, took their toll of victims.[19]

Private Polley, a member of Company F, 4th Texas Infantry Regiment, shared his own observations about the destructive winter:

> Much sickness prevailed among the Texans—more, perhaps, than in commands from the Southern Atlantic States and from Tennessee, where the winters were so nearly equal in severity to those of Virginia. Measles and pneumonia caused the death of many brave young men. Diarrhea led the way to the more fatal complaints. At one time there were not exceed-

18. Davis, *op. cit.*, p. 50.

19. Simpson, *op. cit.*, p. 82.

> ing twenty-five men fit for duty in the 5th Texas, although it had in camp fully eight hundred men.[20]

The Confederate government knew that they could not sustain losses of this magnitude and still defeat the Federal forces. It was decided that each company would send an officer and an enlisted man back to their home county and city to recruit privates for the duration of the war. The Texas Brigade set a goal of 1500 men, a number which gives an ominous picture of how heavy their losses were during that first winter. Lieutenant Lambert was selected to represent B Company;[21] the name of the enlisted man that accompanied him is not recorded. Chaplain Davis left two weeks after the others to help in the recruiting efforts.[22] Lambert concentrated his task in and near Austin. His efforts and the efforts of the other recruiters had not been coordinated with the state of Texas and the state authorities were not too happy about the situation.

Somewhere in Lieutenant Lambert's recruiting trips he ran into John Tatum or John went out of his way to find the Lieutenant. According to the 1860 Federal Census, John was working as a blacksmith in Pilot Point, Denton County, a short distance north of Fort Worth and Dallas. In his household at the time were his wife, Margaret, and four children: John Slack, 12; Mary Slack, 10; Margaret Tatum, 2; and, James, 2 months. John's wife had been born in Pennsylvania, John in North Carolina. Before marrying John, Margaret was married to Henry Slack who was also working as a blacksmith in Armagh Township, Mifflin County, Pennsylvania (about thirty miles south of State College). Besides the two Slack children, Margaret's younger sister, Barbara Emerick, was living with them. Henry probably died somewhere between 1851 and 1855. John met Margaret, later married her in Meadville, Pennsylvania, and moved his new family to Texas where they had Margaret in 1858 and James in April, 1860.[23]

20. J. B. Polley, *Hood's Texas Brigade* (New York: Neale Publishing Co., 1908), p. 17.
21. This is based on several notations in John Tatum's war record.
22. Davis, *op. cit.,* p. 52.
23. This information is taken from the 1850 and 1860 Federal Census, and from family history. It is possible that Henry slack and Margaret divorced but due to the high mortality rates at the time it is more likely that Henry died.

Margaret and a new-born child died during childbirth in middle to late 1861 or very early in 1862, leaving John with four small children. Her death, or serious condition, was the most likely reason that John did not initially join Company B prior to the company leaving for Richmond. After his wife's death and in order to join the 4th Texas Infantry, John farmed the children out to various families for care where they awaited their father's return. No information has been handed down as to the families that took in John and Mary Slack and young Margaret. It is possible that John and Mary returned to relatives in Pennsylvania after John's death. Margaret was probably adopted somewhere in central Texas by her new family. The change in her last name complicated further research into her later life.[24]

Marion "Kingfish" Tatum remembered vividly "Aunt Sophronzie," the woman that raised his father James. As a young boy "Kingfish" remembers her as an old woman rocking on the front porch smoking a clay pipe. "Sophronzie" is a nickname for "Sophronia." The McBee's were close friends with John Tatum prior to his departure for Virginia and the war. It is highly probable that "Aunt Sophronzie" is a McBee. In fact, the 1870 Federal Census shows a Jessie and Sophronia McBee living in Atascosa County, Texas, a few miles south of San Antonio. This is the same area where James spent his early years. There is one additional piece of evidence that this is the right family.

The fact that John and Margaret married in Meadville, Pennsylvania, comes from "Slim" Tatum based on information supplied him by his father. He also remembers his father mentioning that his grandmother's maiden name was Emerick. John was born in 1827. Margaret's birthday is less certain, most likely between 1824 and 1827. John Tatum's neighbor in Pilot Point was William and Louisa McBee. Their granddaughter, Louisa, would later become James Tatum's wife. "Kingfish" and "Slim" Tatum recall their father talking about an older sister but they didn't have any further information about her past. James probably grew up knowing little about his older sister. As far as the family knows, James never mentioned his half brother, John Slack, and his half-sister, Mary Slack.

24. "Kingfish" and "Slim" Tatum recall their father, James, talking about his mother dying during childbirth just prior to, or at the beginning of the Civil War.

The 1870 Federal Census shows several children living with Jessie and Sophronia McBee, including a boy—aged 10—by the name of James. James would have been 10 by the time of the 1870 census. Although the record does not show James' last name as Tatum, it would not be unusual for the census to show all the children in the household as being part of the McBee family. The census also indicates that Sophronia would have been an old woman when "Kingfish" was a boy.

John Tatum first appears on the roster of Company B on March 15, 1862. This same record indicates that Lieutenant Lambert recruited John in Austin, Texas. The Lieutenant, with his new recruits, arrived back in Virginia sometime in late April. While the Lieutenant was gone, the Texas brigade broke camp on March 8 and marched toward Fredericksburg. Colonel Wigfall, just before leaving their winter quarters, resigned his position to become a senator in the Confederate senate, representing the states of Texas. On March 10, Colonel Hood was promoted to Brigadier General and moved up to command of the brigade. At the same time Lieutenant Colonel Marshall was promoted to colonel and to command of the 4th Texas.

The morning of their departure, Colonel, soon to be Brigadier General, Hood delivered this speech to the 4th Texas:

> Soldiers—I had hoped that when we left our winter quarters, it would be to move forward; but those who have better opportunities of judging than we have, order otherwise. You must not regard it as a disgrace—it is never a disgrace to retreat when the welfare of your country requires such a movement. Ours is the last brigade to leave the lines of the Potomac. Upon us devolves the duties of a rear guard, and in order to discharge them faithfully, every man must be in his place, at all times. You are now leaving your comfortable winter quarters to enter upon a stirring campaign—a campaign which will be filled with blood, and fraught with the destinies of our young confederacy. Its success or failure rests upon the soldiers of the South. They are equal to the emergency. I feel no hesitation in predicting that you, at

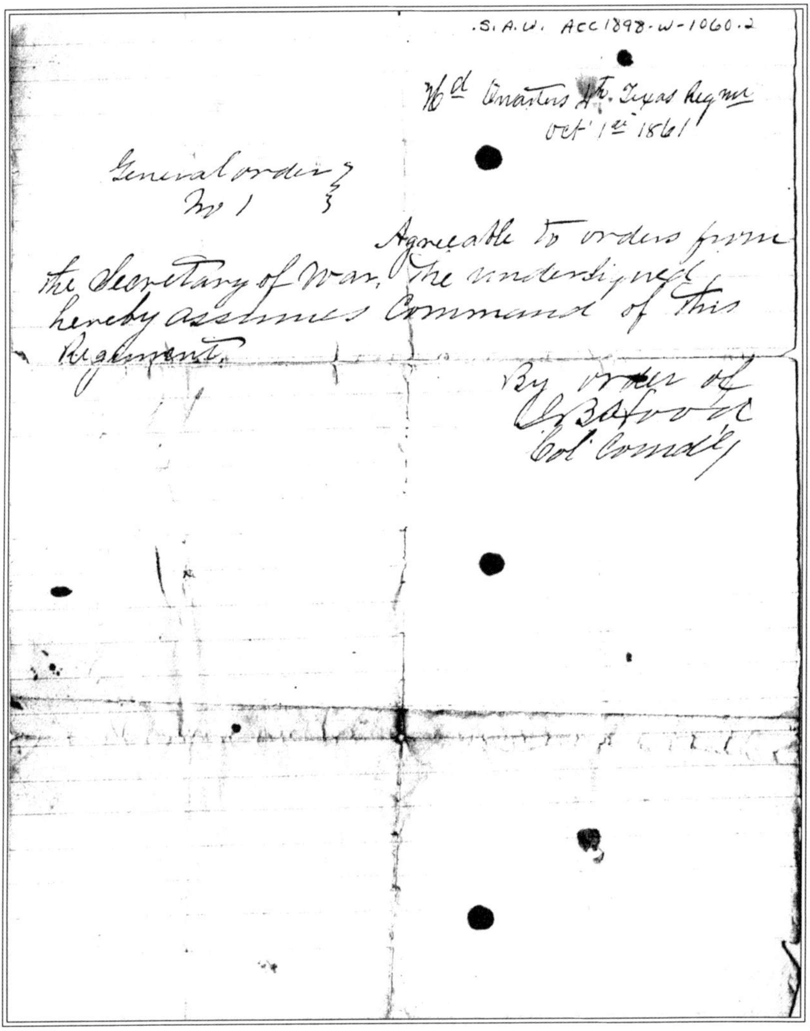

.S.A.W. ACC 1898-W-1060.2

Hd Quarters 4th Texas Regmt
Oct 1st 1861

General Orders
No 1

Agreeable to orders from the Secretary of War, the undersigned hereby assumes Command of this Regiment.

By order of
J. B. Hood
Col Comdg

The order reads, "Headquarters 4th Texas Regiment, Oct 1st 1861, General Orders No. 1, Agreeable to orders from the Secretary of War, the undersigned hereby assumes command of this regiment. By order of J. B. Hood, Col Commanding." The original of this document was discovered by the author in a cardboard box at the Military History Institute, Carlisle Barracks, Pennsylvania, in 1987. It has never been published until now.

least, will discharge your duties, and when the struggle does come, that proud banner you bear, placed by the hand of beauty in the keeping of the brave, will ever be found in the thickest of the fray. Fellow-soldiers—Texans—let us stand or fall together. I have done."[25]

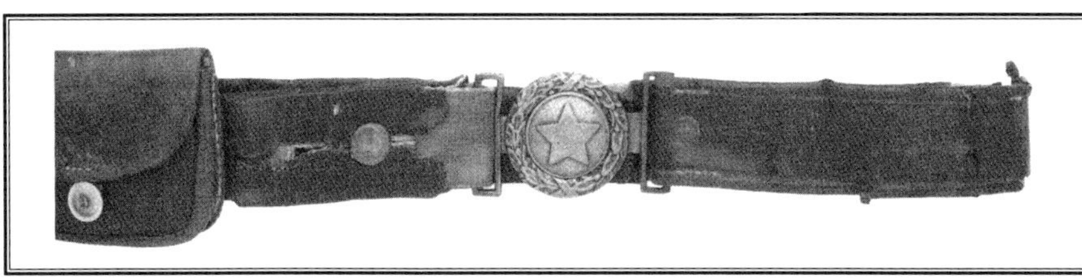

Texas officer's sword belt

25. Davis, *op. cit.*, p. 53.

Chapter Two

Eltham's Landing

Brigadier John Bell Hood now had command of the entire Texas Brigade at the ripe old age of thirty. The Eighteenth Georgia Infantry and Hampton's Legion from South Carolina were added to the brigade to form a brigade of five infantry regiments along with Reilly's North Carolina Artillery Battery. Many consider Reilly's unit to be the best, and perhaps the most famous, battery serving in the Civil War. The brigade later added two additional batteries from South Carolina. Hood's brigade was one of the two brigades in Whiting's division. It would not be long before this would be known as "Hood's Division." The division had 4,320 men while Hood's brigade had 1,922 men on its roster.[1]

Plans were being made in the north for another march on Richmond. The first effort had failed in a miserable defeat at Manassas Junction, also known as the first battle of Bull Run. Lincoln gave control of the army to George B. McClellan with a generous supply of men and supplies. McClellan spent the winter of 1861-62 drilling his men and preparing them for battle. The slow pace of his progress constantly brought grief to the president. Lincoln finally lost patience and offered McClellan two avenues for attacking the Southern capitol at Richmond: a direct march over land, or a flanking movement via the York Peninsula starting at Fort Monroe. McClellan chose the latter.

1. John P. Dyer, *The Gallant Hood* (Indianapolis: Bobbs-Merrill Co., 1950), p. 62.

McClellan landed 90,000 soldiers near Fort Monroe in April, 1962, but this was not the only threat to the new confederacy. McDowell's corps of 40,000 men were operating near Fredericksburg, Bank's corps with 20,000 troops near Winchester, and Fremont was moving into the upper Shenandoah Valley with 15,000. Against this force of 165,000 men, the Confederacy could muster approximately 80,000 men. The South would need a sound strategy to defeat or neutralize a threat of this size. Lee advised President Davis that Confederate forces delay McClellan while Jackson moved up the Shenandoah Valley and threatened Washington, causing McDowell, Fremont and Banks to remain in place and eliminating the possibility of reinforcements being sent to McClellan. This strategy would dictate Hood's movements between March and June.

Hood's brigade marched from their winter quarters near Dumfries, Virginia, to Fredericksburg where they remained for a month. In early April they moved by train to Richmond then marched to Yorktown. This is the same Yorktown that played a key role in the Revolutionary War. This was the longest march the Texans had experienced at that point in the war, some sixty miles over seven days on difficult roads. The Southern forces had established a defensive line from Yorktown across the peninsula to the James River, a distance of fifteen miles. A second defensive line was established up the James River near Williamsburg. Magruder commanded the lines near Yorktown while Longstreet commanded the defense further up the James. These defensive positions were not impregnable, but they served to delay McClellan while Jackson was making his feint up the valley toward Washington.

McClellan had a strategy that looked good on paper. He designed a frontal attack on Yorktown while McDowell's army would go by boat up the York River to Eltham's landing or West Point. McDowell would march west across the peninsula toward the James and cut the Confederate forces in half. It wasn't a bad strategy; the only problem, it didn't work. Several of McClellan's forces ran into opposition that was so strong that plans had to be changed. The Confederate guns along part of the York River kept the Federal gun boats at a distance and immobilized. To make matters worse, McClellan received word from Washington that McDowell's army was being withdrawn from his army to help protect Washington. McClellan was not happy and pleaded with the president to give him Franklin's division. He got his wish and the division was sent by boat up the York to West Point. Franklin finally had his division in place on May 4.[2]

In response to McClellan's earlier moves, Hood was ordered to move his brigade south, then east. The brigade marched from Fredericksburg to Milford Station during a three day period of terrible late-winter storms. From there they boarded a train for Ashland and then marched the eighty-five miles to Yorktown. The brigade made camp a few miles from Yorktown on the same ground occupied by the United States forces eighty years earlier during the Revolutionary War. The Texans had little to occupy their time after their arrival with the exception of sending sharpshooters to harass the Federal lines. It was also during this time that the boys of the 4th Texas gave a new horse to Brigadier Hood.[3]

Eventually General Johnston had to abandon the defensive lines near Yorktown and the James River. The Federals had flanked the Confederates with their gun boats on the York and James Rivers. In addition and even more critical, they had moved large siege guns into place on the lower peninsula. These guns were capable of taking a very heavy toll on Johnston's army. The plan was to move the Confederate forces back to defensive lines a few miles outside of Richmond. This would provide a much stronger defensive position and one that was easier to support. The Texas brigade was given the honored position of being the rear guard. It was their job to hold off the advancing Federals while the rest of the army moved to their new defensive positions. Fortunately the task proved to be fairly easy.

While Whiting's division, and along with it, Hood's brigade were moving past Williamsburg, they received word that Franklin's divi-

2. Dyer, *op. cit.*, p. 65. McClellan's move on Richmond is known to Civil War historians as the Peninsula Campaign.

3. Harold B. Simpson, *Hood's Texas Brigade: Lee's Grenadier Guard* (Dallas: Alcor Publishing Company, 1983), p. 96.

sion was about to land near Eltham's Landing with the mission of destroying Johnston's supply lines and his artillery. The Texans marched toward this location on May 4 and 5. On May 6, Texas scouts were sent forward to determine Franklin's strength and location. The next day Hood led his brigade into battle for the first time.

The military records do not indicate when John Tatum finally connected with Company B, 4th Texas Infantry Regiment. His trip from Texas would have started in late April. It would have been possible for him to arrive in time to participate in the battle at Eltham's Landing. His military records indicate with certainty that he participated in the battle at Gaines Mill which took place one month later. His arrival in Virginia would not have been later than mid May.

Typical of Civil War tactics, Hood led the advance forward with the 4th Texas as the lead element. It was during this march into enemy territory that Hood almost lost his life. Hood describes the event:

> On the 7th, at the head of my command, I proceeded in the direction of Eltham's, with the intention to halt and load the muskets upon our arrival at the cavalry outpost. I soon reached the rear of a small cabin upon the crest of the hill, where I found one of our cavalrymen half asleep. The head of the column, marching by the right flank, with the 4th Texas in the front, was not more than twenty or thirty paces in my rear. When, simultaneously with my arrival at the station of this cavalry picket, a skirmish line, supported by a large body of the enemy, met me face to face. The slope from the cabin toward the York river was abrupt, and, consequently, I did not discover the Federals till we were almost close enough to shake hands. Leaped from my horse, ran to the head of my column, then about fifteen paces in rear, gave the command, forward into line, and ordered the men to load. The Federals immediately opened fire, but halted as they perceived our long line in rear. Meanwhile, a corporal of the enemy drew down his musket upon me as I stood in front of my line. John Deal, a private in Company "A," 4th Texas Regiment, and who now resides in Gonzales, Texas, had fortunately, in this instance, but contrary to orders, charged his rifle before leaving

> camp; he instantly killed the corporal, who fell within a few feet of me.[4]

Chaplain Davis gives a general description of the action of Company B along with the rest of the brigade:

> The command was put in motion at daylight of May 7th, and about 7 o'clock AM, came upon a picket of the enemy, who fired two shots at Gen. Hood, who was riding at the head of the 4th Texas, now in front. One shot struck Corporal Sapp, of Co. H, in the head, inflicting a severe but not dangerous wound. Private John Deal, of Co. A, whose gun was loaded, immediately fired upon the pickets as they ran, and struck the only one in sight, killing him instantly. Some confusion was observed at first in consequence of empty guns, and Col. Marshall's order to "Fall back into the woods and load;" but Gen. Hood immediately called out to the men to "move up," which they did at double quick, and line of battle was immediately formed on the brow of a hill. Beyond this hill, which had a precipitous descent, was an open field of six or eight hundred yards width. On the opposite side were some four or five companies of the enemy, who immediately began falling back into the timber, but not until several random shots had been fired by our men, which we afterwards discovered had killed five and wounded as many more. Company B (Capt. Carter) was then ordered by Gen. Hood to deploy as skirmishers and "feel the enemy." They advanced across the open field, and entering the timber, began a "running fight." Co. G (Captain Hutcheson) was then ordered forward to support Co. B, if necessary; if not, to deploy on its right-the latter course was adopted. Co. K. (Capt. Martin) was next sent to support Co. B, and Co. E (Captain Ryan) to the support of Co. G. After retreating about half a mile, the Yankees made a stand behind an old mill-dam, and a spirited engagement ensued between them and the right platoon of Co. B; under Captain Carter, and Co. G, Capt. Hutcheson. Co. H (Capt. Porter) now arrived upon the ground, with orders to support the left platoon of Co. B, under Lieut. Walsh. The firing now became general, and the enemy, many of their guns missing

4. John Bell Hood, *Advance and Retreat,* (New Orleans: P. G. T. Beauregard, 1880), p. 21.

fire, threw them down and fled. While pursuing them, the second platoon of Co. B came upon a large force (some two hundred,) protected by a heavy palisade. This was more than was bargained for, and the boys, some twenty-five in number, immediately *"treed,"* and answered their volleys, by picking off everyone who showed his head. At this juncture Gen. Hood appeared, and ordered the Lieutenant in command to charge the works, and he would send support. Just as the command "charge" was given, and the boys with a yell, had started for the works, the first platoon of Co. B appeared upon the left flank of the palisade, and the Yankees fled in confusion, leaving seventeen killed and several wounded in the track of their flight. While Co. B was thus engaged, Co. G had also had its share of "fun." Discovering a company of about eighty Yankees, Capt. Hutcheson with his company and part of Co. E, attacked them so vigorously, that they dared not run, and were so unnerved, that they fired volley after volley into the tree-tops. Capt. Hutcheson, who was a Chesterfield in manner, did not for a moment forget himself during the fight. "Charge them, gentlemen, charge them." "Aim low, gentlemen, aim at their waistbands," were his constant exhortations, until a portion of the enemy cried for quarters. "Throw down your arms, gentlemen, you scoundrels, throw them down." Sixteen obeyed the order, and the remainder taking advantage of the momentary cessation of hostilities, turned and fled. Bewildered, however, they took the wrong direction, and coming upon the 5th Texas where it was lying down in line-of-battle, they were greeted by a volley, which left *not one* standing. The fruits of Capt. H's victory, were eleven killed, several wounded, and sixteen prisoners, together with several stand of arms. While these events were transpiring, the 1st, 5th, and remainder of the 4th Texas had entered the timber, leaving the 18th Georgia to support the artillery in the rear. A Yankee regiment now appeared upon the left and rear of the skirmishers, with the intention, doubtless, of cutting them off. Here we witnessed for the first time, the gallantry of the 1st Texas.

The regiment now advancing, 1st California, evidently *intended* to fight well, and advanced steadily within eighty paces of the 1st Texas, when they halted, poured in a volley, and with three *huzzahs,* attempted to charge. This was expected, and

> "aim low, fire," was ordered by Colonel [A. T.] Rainey, and a discharge followed that seemed to mow down the whole front rank, and sent the remainder in confusion back again. A whole-souled hearty *yell* now went up from the Texans, such as only Southerners can give, and they in turn, charged. But the Californians were not yet ready to yield, and rallying, they made a stubborn resistance, and for about twenty minutes, the fire raged with terrible fury. The Texans charged again, and the enemy broke and fled, leaving about two hundred killed and wounded on the field, and several prisoners in our hands. The loss of the 1st Texas in this engagement, was eleven [six] killed and twenty-one wounded. Among the former, however, we regret to chronicle Lieutenant-Colonel [Harvey H.] Black and Captain Decatur, who were loved and mourned by all, as brave men.
>
> After the rout of this regiment, the enemy did not again attack us, but contented themselves with shelling us from their gunboats, and sweeping the woods with grape, from a battery they had planted upon the river bank, without, however, doing us the slightest injury.[5]

Private Giles, Company B, gives his own account as he saw the fight from the company level. He also shares some insights that show how the boys from Texas acted and reacted throughout the war.

> General Franklin had landed his transports and disembarked a large body of infantry. The woods were dense and we worked our way through the underbrush and vines until we ran up against a line of Federal infantry. When they began firing on us, Lieutenant Walsh gave the order "Lie down," and every fellow dropped to the ground and the bullets passed over us, but did some damage to the main body of our troops behind us. The battle was now on, and B and C companies between the two firing lines.
>
> The Federal gunboats in York River opened with grapeshot and canister, making the old woods howl. We were now in a very unpleasant position between the two fighting lines, and

5. Chaplain Nicholas A. Davis, *Chaplain Davis and Hood's Texas Brigade,* Donald E. Everett ed. (San Antonio: Principia Press of Trinity University, 1962), pp. 59-61.

realizing this fact, Walsh shouted, "By the right flanks," but by that time our main line came up and the enemy was falling back toward the river. They made a second stand, and for twenty minutes that was one of the loudest little fights I was in during the war. It may be that I thought it so noisy because it was my first real battle. We had experienced a good deal of unpleasant picket fighting at Dam No. 1 and No. 2 on the Little Warwick River, while at Yorktown, but the battle of Eltham's Landing was our initiatory battle. It was a mere skirmish compared to some we were in later on.

Nevertheless, Companies B and G of the 4th Regiment went down in that wilderness ahead of the brigade and brought on the fight. We never had a man killed.

If my memory serves me right, the 1st Texas had twenty-eight men killed and wounded. Lieutenant Colonel H. H. Black was mortally wounded, and Captain Decater and ten men were killed on the field. Captain Denney, Commissary of the 5th Texas, was also killed early in the action. The Federal report shows that their loss was more than two hundred men.

Just before sundown we were withdrawn from the dismal woods and went into camp back on the hills a few miles from the battlefield. That night I was detailed for camp guard. A strong picket line had been established between our camp and the enemy. I had been on post only a short time when an ambulance came slowly up the narrow muddy road. I halted the driver and asked him where he was going.

He replied, "I am trying to find the field hospital. I have a wounded officer in my ambulance."

I asked him who it was, and he said, "I don't know."

I took his lantern and looked in the ambulance, where I saw a slight, delicate-looking man with a gray blanket thrown over his lower limbs. His head was resting on a knapsack, his eyes were closed and his face very pale. He muttered something that I did not understand. I handed the lantern back to the driver and told him to move one. It was Lieutenant Colonel H. H. Black of the 1st Texas Regiment, who had been mortally wounded in the battle. He died in the ambulance that night and was buried in the piney woods, near the old road. I don't suppose there is a soldier living today that could point out that lonely grave.

Next morning, May the 8th, we struck in the direction of Richmond. The narrow country roads that we had been traveling were bad enough, but when we struck the Richmond and Williamsburg road in the wake of thousands of Johnston's men, it was simply indescribable.

We had to wade through soft mud up to our knees, and some of the language used by the soldiers on that occasion would not look well in print. At one place, after passing through a sea of yellow mud, we were ordered to halt and rest. That was the first time I ever saw General Joseph E. Johnston. General Hood was sitting on his horse near by, talking to some officers, when General Johnston, with his escort, rode up.

He greeted Hood pleasantly and said, "General, I sent you up to West Point to check the enemy, but not to bring on a general engagement."

"Well, General," said Hood, "those people were so saucy that the boys wanted to give them a little thrashing, and I let them do it."

A few more remarks were passed between the two officers, then General Johnston rode on in the direction of Richmond.

I remember just how he looked on the occasion. He wore a gray cloak with red trimming, the left half of the cloak thrown over his shoulder. His uniform coat was buttoned up to his chin, and a little red rose was pinned on his left breast, and he looked every inch the soldier that he was.

When we reached the Chickahominy River, it was eleven o'clock at night and dark as pitch. A miserable old rickety bridge spanned the sluggish stream. By this time the mud in the narrow road had been worked up to a thinness that would have floated a canoe. The men were mad, tired, hungry, and in no humor either to receive or to obey orders.

General Whiting, who commanded my division at the time, had stationed himself on the east side of the old bridge, and as the men came up through the mud and slush he kept singing out, "Close up, men. Move lively." Some fellow in the ranks called out, "Oh, close your bread-trap, and give us a rest."

The general shouted angrily: "Do you know who you are talking to? This is General Whiting."

"General Whiting be damned. You are nothing but some long-tongued courier," came from the dark, muddy road.

"Officers, arrest that man," yelled the General, but nobody paid any attention to the order, and with an oath the general rode across the bridge and disappeared in the darkness.[6]

Tower Enfield Rifle
Used by the North and the South

[6] Mary Lasswell ed., *Rags and Hope: The Memoirs of Val C. Giles, Four Years with Hood's Brigade Fourth Texas Infantry, 1861-1865* (News York: Coward-McCann, Inc., 1951), pp. 97-99. The insult thrown at General Whiting probably came from one of the Texans in John Tatum's company.

Chapter Three

A Trying Trip to the Hospital

Again Hood's brigade was given the honored position of being the rear guard as the Confederate army re-positioned their defensive lines just beyond the Chickahominy River. The Texans had to turn around and face the Federals several times as they covered the move toward Richmond. Chaplain Davis reports that the Texans crossed the river late on May 13, the very last unit to do so.[1] A few days later the brigade moved to a more permanent location on "Pine Island" on the Mechanicsville Turnpike, a few miles northeast of Richmond.

The brigade made the move to Pine Island on May 17; John Tatum was hospitalized with dysentery on May 16, about two weeks after he joined Company B. He was sent to Chimborazo Hospital in Richmond with a serious case of dysentery. This ailment was common throughout the brigade. The water was poor to almost undrinkable; meat often went bad, sanitary conditions were way below army standards—a common fault for General Hood.

Colonel Simpson gives an example of the intestinal problems faced by the Texans:

> Apparently first engagements sometimes have a salutary effect on soldiers. At least, William Andrew Fletcher and several of his comrades of Company F of the 5th Texas, found

1. Chaplain Nicholas A. Davis, *Chaplain Davis and Hood's Texas Brigade,* Donald E. Everett ed. (San Antonio: Principia Press of Trinity University, 1962), p. 64.

> their initial combat experience so. According to Fletcher, he and a number of men in his company on the day of Eltham's Landing "were sufferers from Camp diarrhea . . . and up to that time had found no cure. So," continued Fletcher, "entering battle, I had quite a great fear that something disgraceful might happen and it was somewhat uppermost in my mind; but to my surprise the excitement, or something else, had effected a cure." Upon questioning his fellow sufferers, the 5th Texas Infantryman learned that they too had become "solidified" in the process of battle action.[2]

Unfortunately John Tatum found no such cure. He was sent, along with thousands of others throughout the war, to Chimborazo hospital in a much weakened condition. Chimborazo was the largest military hospital of its time. In fact, US Army historians have concluded that it was the largest military hospital to ever exist.[3] It was not the kind of experience that any soldier would desire to repeat. Chaplain Davis gives one of the best pictures of the hospital and the poor treatment that John would have faced.

> Chimborazo that charnal house of living sufferers . . . is situated at the extreme eastern portion of the City—too remote for many of them [the Ladies of Richmond] to visit even if circumstances had made it otherwise agreeable; which in all candor and honesty I must say in behalf of the ladies and in vindication of the wrongs of more than ten thousand of sick and suffering men was not the case. The place was so foul and the wards so filthy. The bed so lousy and the clothes so dirty, and the clothes of many of the men so nasty and many others so naked, that it was unreasonable to expect the ladies to minister their kind attentions there. And when I attempted to remove our men to other hospitals the Surgeons seemed offended. And why I could [not] tell; unless thereby the other hospitals would see the manner in which patients were treated under their care. There some of the Surgeons did their duty

2. Harold B. Simpson, *Hood's Texas Brigade: Lee's Grenadier Guard* (Dallas: Alcor Publishing Company, 1983), p. 101. Simpson quotes from Fletcher, *Rebel Private*, p. 16.

3. Based on the author's personal discussions with the military historians at the US Army War College.

and some of their wards were under the circumstances well kept. But this was not the case with all—no not one half. And that the reader may have an idea of the place I will describe it. A plat of ground layed out from east to west and from north to south in regular blocks and streets, the blocks each has a ward built upon it about 100 feet long and 30 feet wide. The streets from north to south is about 25 feet wide between each ward and the streets from E to W are about 50 feet. There are about 100 of these wards on 5 or 6 acres of ground. Each ward will contain about 75 beds, now multiply 100 by 75 and you have 7,500 men thus crowded together and then take into the account the privys under the hill, all along the N and S sides so that when the wind blows from either direction, which during the summer is not very infrequent, and then imagine the chambers carried a little below the brow of the hill and emptied. Together with the stench of several thousand wounds you can imagine how sweet [the] place where Lt. Wood, Hon Roundsavall, Johnny Young, J. T. park, D. M. McColister, G. B. Scott, S. T. Owen, George Daugherty died, after having offered their best efforts and their lives upon the altar of their country['s] liberty. Burring shame will follow the men in authority at Chimborazo while the present generation lives. For the men who suffered will not forget. . . . In view of the neglect and mistreatment of our men I did all that I could to remedy the evils by visiting the hospitals daily—which caused me to walk from 8 to 12 ms per day and inquiring into their wants—as all who have survived their wounds will bear witness—and as far as possible afford the best means I could to render them comfortable—and have become so familiar with the whole sanitary program in and about the City I feel positive in the opinion that official dignity with surgeons and stewards and laziness of avocation with cooks and messes have resulted in the death of hundreds of as good soldiers as ever answered the roll call in the army and as good men as lives in the South. To give the reader an idea I will further state that it was not uncommon to see wounds that had not been dressed for from three to five days at a time. And then when the bandages were revoked it was not unfrequent to find the wound filled with living creatures. This will show to some extent the kind attention of both Surgeon and nurse. Many of

Scale model of Chimborazo Hospital located in the museum on the hospital grounds
National Park Service

> the men told me that they had a little cup of sut black coffee, tong[ue] and sour bread and beef or bacon as the case might be and that frequently cooked the day before, as their only diet for days together and this will give you a view of the loving kindness of Stewards and cooks. Mr. Brundreidge of N. Ala. who waited on a wounded son there had a better opportunity to observe the conduct of the nurses and he told me it was not uncommon when a patient asked for anything—a cup of water for instance—to hear the nurse curse him. Or if he were a Negro and the man wanted a piece of meat or bread he had to be hired to bring and sometimes paid in advance.[4]

Civil War conditions created a perfect environment for dysentery and diarrhea to thrive. Men lived crowded together; ate poor diets of fried meat, bread, and coffee; used the same pan to cook their meal that they used to wash up; and went to the latrine upstream from their camp. Bowel disorders were the most prevalent illnesses on both sides of the Civil War and they killed more men than battle. Dysentery and diarrhea, called *"quickstep"* by soldiers, and *"alvine flux"* by the doctors, with dysentery being distinguished by blood in the

4. Davis, *op. cit.*, pp. 103-105.

Chimborazo Hospital, 1862
US Army Military History Institute

stool. Doctors knew neither how soldiers contracted the condition nor how the diseases should be treated.

The number of soldiers who died from loose bowels is staggering. According to Union records of 1,739,135 cases, 57,265 Yankee soldiers died of dysentery or diarrhea, compared with 44,238 men dying in battle. Sometimes regiments had three-quarters of their men stricken at one time. Usually there was one sick soldier for every four well ones except in summer months, when more suffered.

The Confederate Army of the Potomac reported 36,572 cases of bowel disorders in the first nine months of the war. At Chimborazo Hospital in Richmond, one out of every 10 of the diarrhea and dysentery patients died. Andersonville Prison in Georgia sometimes had 130 men die daily from the disorders.

Treatments to open or close the bowels varied from home remedies of tea made from dogwood bark to outrageous prescriptions. One pamphlet advised, *"Let your beard grow so as to protect the throat and lungs."* Many doctors treated dysentery with opium, and diarrhea with "blue mass", which was a mixture of chalk and mercury. Other treatments included: strychnine, castor oil, laudanum, camphor, turpentine, calomel, lead acetate, silver nitrate, quinine, whiskey, ipecac, and even cauterization of the anal opening.[5] It is likely that John Tatum was treated with opium which, according to modern physicians,

5. Based on historical records maintained by the City of Richmond.

does have a "binding" effect on the bowels. John's military records do not indicate how long he was hospitalized. There is some evidence that he was present for duty most of the month of June which would indicate a stay of two to three weeks.

Civil War Pocket Colt

Chapter Four

Gaines Mill

Word had reached Richmond that McDowell was on his way to join McClellan. The time had come for the Confederates to attack before McDowell moved his forces any closer. A plan was quickly put together but General Johnston received news from Jeb Stuart before the plan could be put into place that McDowell had halted his advance. Apparently Jackson's successful attacks near Winchester put fear in the heart of the War Department. The news meant an adjustment to the original plan. Longstreet would attack the Federal center at Seven Pines on the morning of May 31. Whiting's division with Hood's brigade would move down Nine Mile Road and strike the flank and rear. Hood and his men were waiting for Longstreet to start the battle but the sounds of artillery were not heard until four in the afternoon. Hood, Whiting, and Pettigrew's brigades were moving as quickly as they could to support Longstreet. Hood's brigade outran the other two through the mud. The other two brigades had to turn around and support other elements that were under attack. Hood did not receive this information and continued his march toward Longstreet and the main battle. By the time Hood arrived with his regiments, the sun had set and the main battle had ended for the day.

The rest of Whiting's division found themselves involved in a fierce battle at Fair Oaks in support of General Hatton and General Hampton. The losses by the Confederates were heavy: General Hatton was killed, General Pettigrew seriously wounded and taken prisoner, General Hampton shot in the foot. A total of 1,283 Confederates were killed, wounded or missing. Fair Oaks was considered a Federal victory while military historians would give the Confederates a very slight edge at Seven Pines. At Seven Pines the Federals had 790 killed, 3,594 wounded, 647 missing—a total of 5,031. The Confederate losses were even heavier: 980 killed, 4,749 wounded and 405 missing—a total of 6,134. General Johnston was seriously wounded and had to relinquish his command.[1] No one knew at the time that Johnston's wounds would change the whole course of the war. President Davis had to find a new commander of his army. General Lee, his military advisor at the time, was the only logical choice.

Lee didn't lose any time in implementing a new plan for defeating McClellan. Lee had the uncanny ability of getting inside of McClellan's head. He estimated that McClellan would spend some time licking his wounds before he made any major movements. Lee seemed certain that when that movement came, it would be done slowly and with great caution. Lee was right. There was just enough time to throw up fortifications around Richmond so that any attack could be defended with a minimal force. Once completed, Lee could attack McClellan and drive him back. Every spare man was put to work building breastworks just outside of the city while Lee prepared to strike McClellan's forces north of the Chickahominy River. At the same time, Lee planned to use Jackson's forces to strike the Federal right flank.

While the Confederates were waiting for the next campaign to begin, Jackson asked Lee for reinforcements. Whiting's division which included Hood's brigade was sent with considerable fanfare toward the Shenandoah Valley. They marched through the heart of Richmond where the Texans were always considered a sight to behold. Word spread through Yankee spies in the city that Jackson was being reinforced for a major attack in the direction of Washington. The Texans marched to Lynchburg, rested two days, then boarded a train for Charlottesville. They transferred to another train for their fi-

1. John P. Dyer, *The Gallant Hood* (Indianapolis: Bobbs-Merrill Co., 1950), p. 76.

nal leg to Staunton in the heart of the Valley. General Whiting reported to Longstreet and asked for details concerning the general's plan. No details were forthcoming. Longstreet was keeping everything very secretive and promised Whiting and Hood that they would be told which way to march when the time came. The soldiers were told to not even ask the name of the villages through which they passed. They were not to talk to strangers or civilians and if they were asked any questions, they should answer, "I don't know."

The whole secretive affair became a joke with the Texans. They would answer "I don't know" to every question, no matter who asked. Jackson found this out when he encountered a Texan stealing fruit from a tree near the road.

"Where are you going?" asked Jackson.

"I don't know," replied the Texan.

"To what command do you belong?"

"I don't know."

"What state are you from?"

"I don't know."

"What is the meaning of this?" Jackson asked of another Texan nearby.

"Well," he said, "Old Stonewall and General Hood gave orders yesterday that we were not to know anything until after the next fight."

Jackson laughed and went his way.[2]

There was no better officer that served under Lee than Stonewall Jackson. He was a devoted Christian, a conservative Presbyterian, and a master at his trade. Whenever possible, he would stop all marches on Sunday and encourage his men to attend services. His wife continued to run a Sunday School class for black children in their neighborhood while Jackson was away. It was a neighborhood group Jackson had started in his home before the war and one he

2. J. B. Polley, *Hood's Texas Brigade* (New York: Neale Publishing Co., 1910), p. 35.

supported by sending to his wife a portion of his military pay. Unfortunately, Lee lost his best general after the battle of Chancellorsville when a Confederate soldier accidently shot and seriously wounded Jackson. Jackson's arm was amputated but he soon died as a result of his injuries.[3]

Jackson purposely waited until Monday morning to issue his marching orders. Hood was told to reverse his course and march back to Charlottesville. Maintaining the utmost secrecy, Jackson would not give him any directions beyond that point. Many of the men thought they were going to move north by rail and attack somewhere near Washington. Hood discovered on reaching Charlottesville that all of Jackson's army was on its way to a point six miles east of Ashland, Virginia, some sixteen miles due north of Richmond. The plan Jackson worked out in consultation with Lee included moving southeast from Ashland in order to attack the Federal right at Beaver Creek Dam. Other units would reinforce Jackson just prior to the attack and assist him in securing several important bridges held by the Federals.

Jackson was running three hours late. Generals D. H. Hill, A. P. Hill, Longstreet and Lee were all anxiously awaiting his arrival. Lee could see the enemy occupying the town of Mechanicsville while reinforcing their position behind Beaver Dam Creek. Jackson was now seven hours late. A. P. Hill decided he could wait no longer and pushed his troops across Meadow Bridge and began a skirmish with the enemy just outside of Mechanicsville.

On the morning of June 26, Jackson ran into Federal pickets who destroyed the bridges to his front. Texas skirmishers crossed over the creeks and successfully drove the Federals back so that the bridges could be repaired. Jackson's entire army bivouacked for the night at Hundley's corner. During the delay, the rest of Lee's army was en-

3. For most of the war, the 1st, 4th, and 5th Texas Infantry Regiments served in Longstreet's corps. Longstreet maintained an impressive record throughout the war, including his role as a valued advisor to General Lee. Longstreet and Hood became good friends. In fact, Hood took over Longstreet's insurance business in New Orleans at the end of the war. Based on personal research by the author in Lexington, Virginia, and at the Virginia Military Institute.

gaged in a desperate battle in Mechanicsville. As a result of four hours of intense battle, the Confederates lost 1,484 men killed and wounded while the Federal losses were 256. The Federals continued to occupy their defensive positions when night fell on the bloody scene.[4]

McClellan, however, knew that he could face further attacks on his right. He found ground nearby that presented an even stronger position. During the night he moved his forces to the east just beyond Boatswain Swamp. Dyer draws a verbal picture of their defensive position:

> Powhite Creek runs north and south, roughly parallel with Beaver Dam Creek. It is a small sluggish stream with marshy banks which in rainy weather became virtually impassable. Behind this natural defense Lee had expected to find the Federal lines running north and south. Instead, he soon found that the enemy was behind Boatswain Swamp a half mile farther east and that his lines were almost directly east and west instead of north and south as expected. McClellan's engineering ability was clearly demonstrated. He had selected the strongest possible defensive position. As one approached he found the topography ideal for the Federals. There was an almost impenetrable growth of trees, then the swamp and the creek, more swamp on the opposite bank and then an incline which led to the top of a low hill. The slope was bristling with two lines of infantry behind log and earth works protected by abates. On the hill were other lines of infantry and batteries of artillery.[5]

General Fitz John Porter had his three divisions, about 25,000 men, manning the defensive lines in this area. The first line of defense near the creek consisted of Companies C and G of the First United States Sharpshooters under Colonel Hiram Berdan, also known as the Berdan Sharpshooters. On the slope behind them were three infantry brigades of George Morell's Division. They had two lines of defense: one half way up the slope, and one near the top of the hill. They were supported by eighteen guns under Captain William Weeden. He was supported by even more infantry. Considering the

4. Dyer, *op. cit.*, p. 85.

5. Dyer, *op. cit.*, p. 86.

terrain, the defensive units and positions, this was as strong a defense as seen anywhere during the Civil War.

Although this sounds formidable enough, Colonel Simpson adds this to the picture:

> The terrain west of Boatswain's Creek over which an attack against Porter's left would have to be made, was ideal for the defense. Although a thin strip of timber was located a quarter of a mile west of the creek, the terrain from this woods down toward the creek was an open and slightly undulating plain some 700 to 800 yards wide which offered little protection to an attacker. Boatswain's Creek itself posed a definite obstacle. The shallow stream bed was about ten feet wide with rather steep banks and marshy approaches that would perceptibly slow down infantry. Into this combined natural and man-made death trap the Texas Brigade would be ordered.[6]

Throughout the 27th A. P. Hill attacked the center of the Federal lines with little effect and with terrible loss of life. D. H. Hill started his attack on the Federal right about 2:00 PM with the same result. Hill threw his regiments toward the creek time after time; every attack was thwarted with heavy losses. Things were not going well and Lee was getting desperate for a victory. Hill's troops were staggering to the rear, demoralized and badly shaken. It was 5:00 in the afternoon and Lee knew he had time for one more effort before the day came to an end. He would use Hood's and Law's brigades as his attacking force with the hope that they would succeed where so many others had failed.

The description of what took place at this point is best given over to those who were there. Everyone, from general to chaplain, had a different view of the battle. We will let them tell their story starting with General Hood.

> We had heard during the day, in the direction of Mechanicsville, the guns of Longstreet and A. P. Hill, which indicated that the issue of the great battle, then in progress, would soon be decided. At early dawn of the 27th the march was resumed;

6. Harold B. Simpson, *Hood's Texas Brigade: Lee's Grenadier Guard* (Dallas: Alcor Publishing Company, 1983), p. 115.

Ewell's Division bore off in the direction of our left during the day, and Whiting's to the right. The latter received instructions, in the afternoon, to repair to the support of Longstreet, then assaulting the Federal left at Cold Harbor. I moved on with all possible speed, through field and forest, in the direction of the firing, and arrived, about 4.30 PM, at a point, on the telegraph road, I should think not far distant from the centre of our attacking force. Here I found General Lee, seated upon his horse. He rode forward to meet me, and, extending his usual greeting, announced to me that 'Our troops had been fighting gallantly, but had not succeeded in dislodging the enemy; he added, "This must be done. Can you break his line?" I replied that I would try. I immediately formed my brigade in line of battle with Hampton's Legion on the left. In front was a dense woods and ugly marsh, which totally concealed the enemy from us; but the terrible roar of artillery and musketry plainly revealed, however, that thousands and thousands of living souls were struggling in most deadly conflict for the mastery of that field, and I might say, almost under the shadow of the Capitol of the infant Confederacy. My line was established, and moved forward, regiment by regiment, when I discovered, as the disposition of the Eighteenth Georgia was completed, an open field a little to its right. Holding in reserve the 4th Texas, I ordered the advance, and galloped into the open field or pasture, from which point I could see, at a distance of about eight hundred yards, the position of the Federals. They were heavily entrenched upon the side of an elevated ridge running a little west and south, and extending to the vicinity of the Chickahominy. At the foot of the slope ran Powhite creek, which stream, together with the abates in front of their works, constituted a formidable obstruction to our approach, whilst batteries, supported by masses of infantry, crowned the crest of the hill in rear, and long range guns were posted upon the south side of the Chickahominy, in readiness to enfilade our advancing columns. The ground from which I made these observations was, however, open the entire distance to their entrenchments. In a moment I determined to advance from that point, to make a strenuous effort to pierce the enemy's fortifications, and, if possible, put him to flight. I therefore marched the 4th Texas by the right flank into this open field, halted and dressed the line whilst under fire of the long-range guns, and gave positive instructions that no man should fire until I gave the order; for I knew full well that if the men were allowed to fire, they would halt to load, break the alignment, and, very likely, never reach the breastworks. I moreover ordered them not only to keep together, but also in line, and announced to them that I would lead them in the charge. Forward march was sounded, and we moved at a rapid, not at a double-quick pace. Meantime, my regiments on the left had advanced some distance to the front through the swamp.

Onward we marched under a constantly increasing shower of shot and shell, whilst to our right could be seen some of our troops making their way to the rear, and others lying beneath a galling fire. Our ranks were thinned at almost every step forward and proportionately to the growing fury of the storm of projectiles. Soon we attained the crest of the bald ridge within about one hundred and fifty yards of the breastworks. Here was concentrated upon us, from batteries in front and flank, a fire of shell and canister, which ploughed through our ranks with deadly effect. Already the gallant Colonel Marshall, together with other brave men, had fallen victim in this bloody onset. At a quickened pace we continued to advance, without firing a shot, down the slope, over a body of our soldiers lying on the ground, to and across Powhite creek, when, amid the fearful roar of musketry ad artillery, I gave the order to fix bayonets, and charge. With a ringing shout we dashed up the steep hill, through the abates, and over the breastworks, upon the very heads of the enemy. The Federals, panic-stricken, rushed precipitately to rear upon the infantry in support of the artillery; suddenly the whole joined in the flight toward the valley beyond. At this juncture some twenty guns, stationed in rear of the Federal line on a hill to my left, opened fire upon the 4th Texas, which changed front, and charged in their direction. I halted in an orchard beyond the works, and dispatched every officer of my staff to the main portion of the brigade in the wood on the left, instructing them to bear the glad tidings that the 4th Texas had pierced the enemy's line, and were moving in his rear, and to deliver orders to push forward with utmost haste. At the same moment I discovered a Federal brigade marching up the slope from the valley beyond, evidently with the purpose to re-establish the line. I ran back to the entrenchments, appealed to some of our troops, who, by this time, had advanced to the breastworks, to come forward and drive off this small body of Federals. They remained, however, motionless. Jenkins's command, if I mistake not, which was further to our right, boldly advanced and put this brigade to rout. Meantime, the long line of blue and steel to the right and left wavered, and, finally, gave way, as the Eighteenth Georgia, the 1st and 5th Texas, and Hampton's Legion gallantly moved forward from right to left, thus completing a grand left wheel of the brigade into the very heart of the enemy. Simultaneously with this movement burst forth a tumultuous shout of victory, which was taken up along the whole Confederate line.

I mounted my horse, rode forward, and found the 4th Texas and Eighteenth Georgia had captured fourteen pieces of artillery, whilst the 5th Texas had charge of a Federal regiment which had surrendered to it. Many were the deeds of valor upon that memorable field. General Jackson, in reference to this onset, says in his official report:

> ... In this charge in which upwards of a thousand men fell, killed and wounded, before the fire of the enemy, and in which fourteen pieces of artillery and nearly a regiment were captured, the 4th Texas, under the lead of General Hood, was the first to pierce these strongholds and seize the guns. Although swept from their defenses by this rapid and almost matchless display of daring and **desperate valor**, the well disciplined Federals continued in retreat to fight with stubborn resistance," [7]

Chaplain Davis had the opportunity to witness part of the battle and to interview many of the soldiers after the attack. He gives a more detailed account of the attack.

> At 4:15 PM we heard the first volley of musketry. We were halted--countermarched by regiment the order having been right in front & fronted in line of battle in the direction of the firing

[7] John Bell Hood, *Advance and Retreat* (New Orleans: P. G. T. Beauregard, 1880), pp. 25-28.

Desperate Valor -Dale Gallon
General Hood making the attack at Gaines Mill with John Tatum to his left with rifle in hand.
Image courtesy of Gallon Historical Art: www.gallon.com

(south) the 5th, 1st, 4th, 18th, Hampton Legion. At the counter marching the 4th became a little confused by order of Col. M. [arshall] in counter marching. When formed we advanced a short distance into a skirt of timber & remained for a short time-waded a slough, halted a short time while moving to the next. Here Col. M. sent his orderly to Lt. Col. W. [arwick] informing him that he was Col. of the Reg. & would comd. it. Lt. Col. W. had said to me 3 hours before that he would command, he also spoke of his serious reflections & said he never expected to survive the fight.

Gen. Lee & Whiting rode up inquiring for Gen. Hood--after a minutes interview with him Gen. Hood rode off in the direction of the fight but in a short time returned on foot, hat in one hand & sword in the other, he had been looking the situation of the enemy & nature of the ground over which he expected to move,-and called out where is my old 4th Tex.? They had layed down in consequence of their exposed situation & were concealed from view by the bushes & timbers fallen by the enemy-"here we are General," was the general response as they sprang to their feet ready to do and die at his command. "I want you to follow me." And then came a shout that made the forest ring. "Forward" was distinctly heard & recognized by every man as the battle shout of their old leader. They sprang forward as if by magic & passed up the hill & emerged into an open field; which when fully gained & then obeying the order "Left Wheel" they were confronting the enemy, yet at a distance of about 600 yds. "Forward." And their whole line struck off at a double quick-"Steady." "Steady." I don't want you to run. Move steadily. Then at a "Quick" steady march they swep[t] in solid line across the field. The General himself and on foot leading them like a champion to the charge. It would be a difficult task here to describe feelings of mutual confidence which existed between this regiment & their general. He had been relieved of the command of cavalry, his favorite service on the Peninsula & appointed to the command & to share the trials and destiny of this Reg. on the______ of Oct. He had taught them the importance of discipline & drill, the manual at arms, in short the service of war. And on the Potomac at West Point & at the Seven Pines he had a sufficient test of their bravery & skill to inspire him with the highest confidence of their ability, in an hour of the greatest peril. Consequently the 4th had been held in reserve to be hurled like a thunderbolt upon the enemy at the point where the fortunes of this awful day were to be determined. The time had come. 40000 men in combat had been struggling for two dreadful hours. Alternate hopes & fears contended with almost every moment of that time. At one time our troops advanced with firm & vigorous step. Again they wavered haulted & withdrew. The muskets roared the artillery thundered-flying horseman & charging squadrons. Advancing legions & retreating divisions presented to the eye of the beholder a scene of wildest confusion. The day was growing dark with the smoke of burning powder & the ear deafened with the hideous thunder of a hundred cannon. The country around was full of wounded soldiers & the ground strewed with expiring warriors. It was 6 o'clock. the sun was fast sinking over the western hills. Light clouds had begun to veil his [God's] face as if to shut out his light from the misfortunes of the struggling heroes & let the mantle of darkness cover the bleeding wounds of expiring freemen. And veil the tears of the goddess of liberty as she wept over the departing hopes of the Sons of the South.

To allow a defeat at this point would give the enemy the power to throw his left wing into Richmond-which he confidently believed he could do as his message to that wing of the army plainly shows. "I have the 'Rebels' right where I want them. So soon as they are checked on the right I will throw the left across the Chickahominy and Richmond falls." This dispatch created the wildest enthusiasm through out that entire wing of the army. And their joy was to be equaled only by their subsequent mortification, at the final & utter defeat of the right wing. The position against which the 4th was to be led seemed to be the key of the whole field. It had been well selected. Naturally strong & the appointment well made-well entrenched & strongly garrisoned. It had resisted several attacks & broken & driven one line after another with its storm of iron hail from the field. And as we advanced to the charge we met two regiments in full retreat and soon after wheeling in the field we passed a branch the channel of which was filled with hiding men & sulking officers. One Reg. had left its battle flag on the field which was picked up by Lt. Col. Warwick & borne on by him as he cheered men towards the goal of his country['s] glory until he was shot down & was covered with his flag-the warriors pride. As they move over the open field there were two lines of infantry & 3 batteries of 24 guns that had them in full sight & under direct range. The rifle's crack, the cannon's roar, the whistly bullets & screaming shells & bursting bombs. The howling balls & whizzing grap[e] thinned & tore our ranks at every step. But steadily & right on they pressed for the enemies lines without halting or delivering a single shot. Four hundred yds. brought them in full view of a third line of infantry which were concealed in a breastwork of logs on the opposite bank of the branch-of steep banks & mirey channel-who opened a deadly fire upon us at a distance of about seventy five yds. Col. Marshall was shot dead and fell from his horse soon after we entered the field and at about 555 yds. from the enemy-taking effect in his neck. Lt. Col. Warwick discovering our near approach to the enemy ordered the men to "Halt" & "Fire." Fortunately General Hood was still with the reg. and instantly cried out "Don't Halt Here." "FORWARD" "FORWARD." And without waiting to reload the men moved at a double quick; loading as they ran, & threw themselves upon the enemy--charging over the mirey branch & over the breastwork put them to flight, and following up the advantage gained pressed the flying garrison upon the second entrenchment, which was so masked by their retreating, cowardly comrades as to protect us from their fire until we had gained near half the distance to the second position. And by the time they were unmasked, seeing Southern Steel advancing at rapid strides, they joined their frightened friends & made for the third line composed of US infantry drawn up on the crest of the hill. The advantage gained at this point was also improved & soon the three lines were mixed in glorious confusion & hastening for shelter towards the heavy battery of 16 guns on the hill to the rear. This battery occupying an elevation above the one occupied by the advanced lines, had been operating fearfully upon us while passing over the first platoon before reaching their first position. And as soon as their retreating lines had passed into the second depression, they opened upon us with frightful terror; Throwing a fearful storm of grape & canister into our midst. Our line had become broken & our men were advancing in some disorder, occasioned by the number of men & officers that had fallen & the irregularity of the ground over which they had to pass & after passing the first breast work & before reaching the top of the hill Lt. Col. W. fell mortally wounded & by his side Capt. E. Ryan also fell, both shot thro' the lungs. Gen. Hood seeing the confusion & that the enemy were advancing up on his right to cut them off, gathered up the scattering men on the right & put them in position to keep them in check and ordered the 4th to fall back to the edge of the timber & "re-form." And at the same time sent off for aid which soon arrived. Putting them in motion again after guarding the prisoners who had been captured about the Yk. hospital, the barn & stable which were filled with skulking Scoundrels to the rear, they were soon protected by the depression in the field from the deadly fire of the last battery on the field. The others having been captured & silenced in the march. Here the 18 Ga. came up & forming with the 4th advanced right in the mouth of their guns. Here as a last & dying struggle there dashed a heavy squadron of cavalry upon our line. But its charge was not more gallant than its retreat was magnificent. There were but few of them who had the courage to reach the sticking point. One man received a bayonet & another carried a musket off in his horse's side. Many saddles were instantly emptied of their riders & many horses fell upon the ground. The line of their retreat was well defined by fallen men & slaughtered horses- And our battle flag was planted upon their guns & midst the shouts of our valiant men. The day is won. The victory is ours and the brave Sons of the South are persuing the defeated far & capturing straggling squads, scattered Cos. & broken Regs. And night at last stops the chase. And with what mingled feeling of pride & sorrow do our brave boys lie dow[n] & rest-for but few can quiet themselves to sleep. Conscious that they have done their duty, and that they have kept the high trust committed to their care-foiled & defeated the invaders of our homes & sent his broken ranks in disorganized confusion flying across the Chickahominy in search of another "base" for future operation which McC[lellan] says he has been considering for some time. But I am persuaded, that had he succeeded in "checking" us at Gains mill & been able to throw his left across the Chick[ahominy] he never would have divulged to the world his intention of "Changing his Base."

Oh what a world of suffering and blood has this evening to record. The sun went down. But the work of slaughter continues. Night comes on. The enemy has fled. The whole field of battle from one end to the other is ours. Thirty six pieces of artillery are captured. Ten thousand stands of arm lie scattered on the field. Abandoned camps are full of supplies that add comfort to a soldiers privations. Quartermaster & commissary stores are found in great profusion & heaped & scattered in great confusion. Knapsacks, Haversack, canteens, overcoats, blankets, shoes, boots, knives, pistols, sabers, bayonets, shovels, picks, axes, pots, kettles, wagons of the ordinance, commissary, transportation & pontoon trains. Ambulances, single, double, Quadruple. Supplies of medicine, in short everything that will tend to fit up a grand invading army are seen in every direction in which the eye is turned. Books of every variety.

Portfolios, papers, letters, journals, diaries, etc., help to make up the sum total of the grand picture of this field of death.

Night having put an end to the pursuit of the flying foe, the long train of ambulances are crowding on the field to gather up the wounded & carry them to the field infirmary. Hundreds have already gone to the rear. But there are hundreds more not able to go. They lie scattered over the field and mark the line of our advancing army. All night long the soldiers who seemed so reckless of friends & of life wanders over the field in search of a dear Bro. or fallen friend. He entered the field with him but since then he has had no Knowledge of him and maybe morning has come before he gets the tidings that he fell as he advanced upon the breastwork or was grappling with the foe at the mouth of the cannon. Maybe his arm was shivered or body pierced with the sabre or a minnie ball & faint from loss of blood & exhausted & famishing for water & food he has fallen by the way & lies yonder in the field or in the woods & his life fast flowing out-he is found. He is dead. None but his God knows the sequel of his career. Whether he instantly expired or if he suffered long hours in excruciating pain. Whether he felt like he was prepared for that solemn change or not. Here lies the warrior upon the ground, Thy battle is fought, Thy victory is won, If my cause succeeds my Countries free, My highest boon is to die for thee, My country Farewell.

After the battle was over Gen. Hood manifested that kind feeling for his men that a long & severe campaign together, is sure to beget. All night he rode & walked to have them gathered & sent to places where they could have their wounds cleansed and dressed. And on the next morning when the line was formed & the roll was called it was found that out of 531 present & entered 261 were missing. And of these 46 were left dead on the field. It was then that the strength of his attachment was seen developing itself in tears over his manly cheeks. And every soldier felt the same emotion. He rode up & inquired "Is this the 4th Tex." Some one answered "this is all that remains." He turned his horse but not soon enough to prevent the regiment from witnessing how well he loved his men. They watched him as he rode using his handkerchief

> until out of sight. There was not a soldier in that line but what thought more of [him] now than ever before.
>
> To the hospitals the wounded were carried all night & all the next day. The yards were soon filled & they were lying all round the yards in every shades & many lay a long time in the sun. This was the most heart rending scene that I have ever witnessed. We worked on thro' the night & thro' the day & next night. And on Sunday morning all having been brought in I left with a train of 13 wagons loaded with the wounded for Richmond. And on arriving they were all sent to Chimborazo.[8]

There were many stories that came out of this battle. The cavalry unit sent by General Porter to protect the Federal retreat was Hood's old unit from his prior days in Texas. The Texans poured a deadly volley into the unit before they could do any damage. Hood's old commander, Captain Charles Whiting, fell dazed at Hood's feet, his horse shot from under him. He was taken prisoner. Captain William Chambliss, another of Hood's old friends, was mortally wounded. Lieutenant Colonel Upton of the 5th Texas, dressed only in his shirt and with a frying pan in one hand, accepted the surrender of the 6th New Jersey. Shortly after the attack and while the 4th was catching their breath, a lone Federal was busy near the stable taking shots at the conquering Texans. Private Jim Stringfield decided to go after him. Leaping the fence, he ran toward the stable. At this point the easy going Lieutenant Hughes of the 4th who did not have much practice at cussing, yelled, "Go it, Stringfield—go it! Kill him, dod damn it, kill him."[9]

The battle took a very heavy toll. At the end of the attack and near the top of the hill, Lieutenant Haskill, General Wade Hampton's son-in-law, ran up to Hood and offered his services as an aide. Hood discovered that the Lieutenant's left arm had been completely shot away. The Lieutenant was so persistent that Hood had to order him to a hospital. Every officer in the 4th Texas above the rank of captain

8. Chaplain Nicholas A. Davis, *Chaplain Davis and Hood's Texas Brigade,* Donald E. Everett ed. (San Antonio: Principia Press of Trinity University, 1962), pp. 86-91.

9. Polley, *op. cit.,* p. 68.

had been killed. In addition, 253 men, almost half of the regiment, were casualties. The total casualties, killed and wounded, in both Hood's and Law's brigades came to 1,016. That evening while Hood was looking after the wounded on the field he heard his name being called at a corner of the field. He sent a soldier to inquire as to source of the call. A soldier reported back that it was Captain Chambliss who was severely wounded during the Federal cavalry attack. The captain wanted to see his old friend. They greeted each other, carried on a short conversation, and then Hood ordered that he be taken to a field hospital to receive the same attention as his own officers.[10]

The 4th Texas, led by General Hood, gave General Lee his very first victory. The Federals were in full retreat. Although other battles followed, the Federals were eventually driven off of the peninsula and back to Washington. The Texas brigade was praised by Lee, Jackson, and Longstreet. Their fame quickly spread through Richmond and in newspaper print throughout Texas. They were rewarded with being placed in the middle of many of the most heated battles the Confederacy would face through the rest of the war.

John Tatum's military record indicates that he took part in the attack. Since there is no record of exactly when he arrived in Virginia, this may have been his first encounter with the Federals, a baptism of fire and destruction. There is no record of him receiving any serious wounds during the battle.[11]

After Gaines Mill, several more bloody battles were fought on the Peninsula: Garnett's and Golding's farms (June 29), Savage's Station (June 29), White Oak Swamp (June 30), and Malvern Hill (July 1).

---

10. Dyer, op. cit., p. 94.

11. Tom Tatum and the author commissioned Dale Gallon, a well-known Civil War artist, to create a painting of General Hood making the attack at Gaines Mill. The painting shows John Tatum near Hood as they cross the first line of Federal infantry. There are no known photographs of John but later photographs show his son James. The artist assumed that John probably looked like his son James. The painting adorns the cover of Brian Burton's book *Extraordinary Circumstances: The Seven Days Battle.* The painting can also be seen at http://www.gallon.com

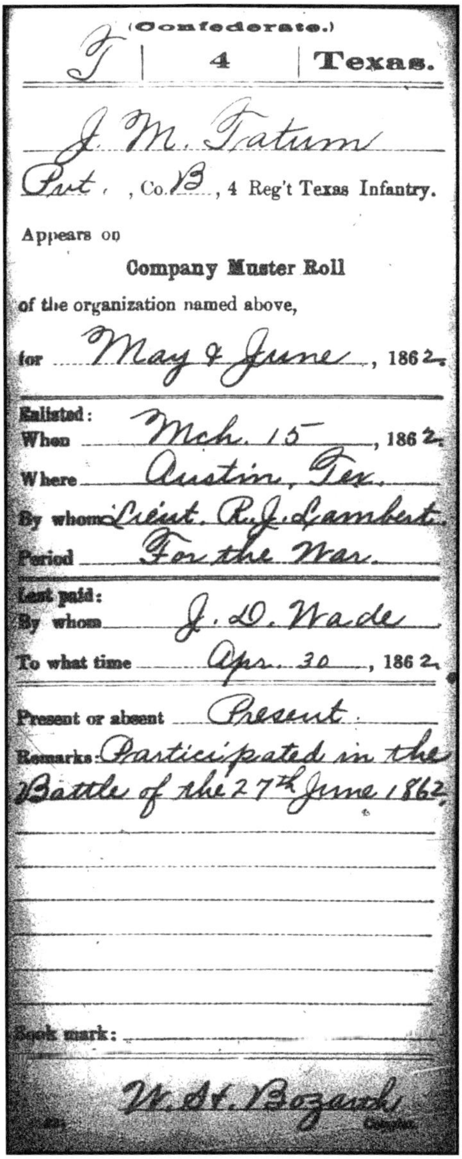
(Confederate.)

T | 4 | Texas.

J. M. Tatum

Pvt., Co. B, 4 Reg't Texas Infantry.

Appears on

Company Muster Roll

of the organization named above,

for May & June, 1862.

Enlisted:
When Mch. 15, 1862.
Where Austin, Tex.
By whom Lieut. R. J. Lambert.
Period For the War.

Last paid:
By whom J. D. Wade
To what time Apr. 30, 1862.

Present or absent Present.

Remarks: Participated in the Battle of the 27th June 1862.

Book mark:

W. H. Bozarth
Copyist.

John Tatum's war record with mention of participation in battle on 27th June 1862.

In these final battles the Federals had far fewer losses but McClellan was convinced that he was on the losing end and had no choice but to retreat and regroup. Hood's brigade and the 4th Texas were not engaged in a major way until Malvern Hill. It was there that Hood wanted to make an attack on the Federal right at a spot he found to be weakly defended. Whiting, his division commander, for some unknown reason would not allow him to advance. Most of the casualties in the brigade were due to artillery fire. At the end of the Peninsula Campaign, also known as the Seven Days Battle, the Confederates could claim a victory, Richmond was saved, but at a severe price: Confederates killed or wounded 19,739; Federals killed or wounded 9,796.[12] Following these series of battles, McClellan pulled his army back to Harrison Landing on the James River in the hope that he would receive reinforcements that would allow him to make another drive on Richmond. Lee knew that his troops were too spent to pursue McClellan any further. He needed time to rest his units and make

12. Dyer, *op. cit.*, p. 103.

General John Bell Hood
(1831-1879)
Library of Congress

some needed changes among his generals. Whiting's division, along with Hood's brigade, were sent a few miles north of Richmond where they camped along the Mechanicsville Road for the better part of a month. During that time Lee managed to move Whiting out of his position and replace him with Hood. From that point on it became known as Hood's division.

Chapter Five

Second Manassas (Bull Run)

Lincoln was hard pressed to find new leadership. This was an issue that plagued him throughout much of the war. With some reluctance he brought in General John Pope from the western theater. The Union was having some success outside of Virginia but the nation was focused on Virginia. Most military leaders felt that the outcome of the war depended on success or failure in Virginia. General Pope had met with limited success on the battlefield, was a known braggart and had a reputation for treating his subordinates with contempt. He was well-connected in Washington and impressed the right people with his aggressive military attitude. Unfortunately, Pope and McClellan were adversaries that served in the same army. Their disagreeable attitudes would not serve Pope well in the upcoming battle.

Pope was asked to form a new army, the Army of Virginia, not to be confused with General Lee's Army of Northern Virginia. It wasn't really new but a re-organization of the Union forces that had been fighting and losing in Northern Virginia. His orders, dated June 26, 1862, gave him command of three corps. Franz Seigel commanded the First Corps. He was a graduate of the German Military Academy. Major General Nathaniel Banks, a general due to his political connections, commanded Second Corps. Irvin McDowell, an experienced soldier and former instructor at West Point, commanded the Third Corps. McDowell also served as Pope's primary advisor throughout the upcoming battle. Pope's general orders were to disrupt the Confederate supply lines in central and western Virginia and to take the offensive in another move on Richmond.

Pope, however, was content with a defensive position near Culpeper where he could harass the Virginia Central Railroad. He saw this as an opportunity to block any move that General Lee might make into northern or western Virginia. This left General Halleck, the commander over the Union armies, with a decision: what to do with McClellan and his army. Halleck was of the old school that said that an army fights best when it is consolidated. McClellan, on the other hand, wanted more troops so that he could renew his advance on Richmond. His argument failed to win the day and he was ordered to send his units by boat to the Fredericksburg area to join Pope's army.

Lee was aware of Pope's arrival but did not know of McClellan's intentions. He could not leave Richmond unprotected nor could he allow Pope to cut off his supply line to the rest of Virginia. He decided to take a risk by sending Jackson and his corps to confront Pope while Longstreet remained behind to protect Richmond. Jackson planned on taking the offensive when the right opportunity presented itself even though his force was considerably smaller. The opportunity presented itself at Cedar Mountain, outside of Culpeper, where Jackson defeated Banks. The defeat took some of the starch out of Pope and increased his need for McClellan's units. McClellan didn't feel the need to rush; the various units from McClellan's army arrived piecemeal and at a very leisurely pace.

Once Lee realized that McClellan was moving off of the Virginia peninsula he moved most of Longstreet's corps to the right of Jackson.[1] Hood's division of two brigades followed closely, stopping for a time at Hanover Junction to ensure the protection of the eastern leg of the Virginia Central railroad.

Pope's forces were camped within the "V" formed by the Rapidan and Rappahannock Rivers, a short distance west of Fredericksburg.

1. Longstreet was a 41 year old graduate of West Point where his classmates included John Pope, and his instructors included Irvin McDowell.

Lee saw this as an opportunity to strike Pope's left flank, cut him off from the units arriving from Fredericksburg, or Washington, and trap him between the two rivers. It didn't take Pope long to realize he was in a bad position. Although he had bragged that he never showed his back to the enemy, he secretly marched his units to the north side of the Rappahannock. Although the retreat caught Lee by surprise, it didn't go without some Union losses.

The 4th Texas, along with the rest of Hood's division, moved north toward Kelly's Ford, when Captain Reilly, the artillery support for the 4th Texas throughout most of the war, got into trouble with General Hood. While both sides were exchanging artillery rounds across the river, a lone Yankee rider entered the river carrying a white flag. "Swearing that although he could see man and horse distinctly," reports Polley, "he could see no flag, Reilly trained and fired a gun at the fellow, the round shot plunging into the water three feet to his right. That not calling him to a halt, Captain Reilly, still insisting that he could see no white flag, fired two more shots." An aide-de-camp dispatched by General Hood came at full speed and shouted, "General Hood says stop your damned foolishness—that man is bearing a flag of truce." "An' so, be Jesus, he is," confessed Reilly, "but in the name of St. Patrick an' all the ither hoully saints, whoy didn't the spalpeen hold the damned white rag high enough for an Irishman to persaive it?"[2]

Further up the river Hood's division moved near Freeman's Ford to relieve Trimble. Bohlen's Union brigade was still on the south side of the river. The Texas brigade attacked Bohlen's left while Trimble moved on his right. Chaplain Davis shares a first-hand account of what followed:

> ... with the line of battle thus formed, the "Forward" was given; the line of enemy was instantly broken, and driven headlong into the river. And pouring a dreadful fire into their crowds of confused and broken lines, as they were huddling together to cross, many were shot in the back, and others drowned by

2. J. B. Polley, *Hood's Texas Brigade* (New York: Neale Publishing Co., 1910), p. 74.

> the crushing crowd which pressed for the other shore. It was the work of but a few minutes, yet about three hundred of their killed and wounded were left upon the ground and in the river. Here Major M. D. Whaley, of the 5th, fell mortally wounded. His thigh was shattered by a shell, and had to be amputated. He died on the next day. We lost only ten men in the fight—all of [from] the 5th Texas.[3]

With Pope in a strong defensive position north of the river, Lee needed a new, bold plan. Military minds at the time would have faulted Lee for taking too big a risk but Lee decided to divide his army, at least for a time. Jackson would take his corps on a long march around Pope's right, across his rear, cut off supplies arriving from Washington, cut off avenues of retreat, and pick a place where Pope could be defeated. Longstreet would continue to defend south of the river, hold Pope in place, and join Jackson later taking the same route as Jackson.

Jackson took his corps across the river at Hinson's Mill Ford, then through the small towns of Orleans, Salem, and While Plains, on through Thoroughfare Gap to Haymarket and Gainesville. He attacked the Orange-Alexandria Railroad at Bristoe Station and destroyed a massive amount of Federal supplies at Manassas Junction. Pope finally discovered Jackson's movements and determined to destroy Jackson near Centreville. Jackson found a better place to establish a defensive position while he awaited Longstreet's arrival, the same place where the first major battle of the Civil War was fought near Manassas. Jackson, knowing the ground from the first Manassas battle, spread his divisions for a mile along a railroad cut that ran between the Warrenton Turnpike (Highway 29) which runs west to east through the battlefield, and Sudley Road (Highway 234) that runs south to north.

Pope was getting frustrated trying to find Jackson. He knew he was in the area; he had the mess in Manassas Junction as evidence.

---

3. Chaplain Nicholas A. Davis, *Chaplain Davis and Hood's Texas Brigade,* Donald E. Everett ed. (San Antonio: Principia Press of Trinity University, 1962), pp. 108-109.

Jackson's location was discovered in an unexpected skirmish on the Warrenton Turnpike. On August 28, King's division was moving east on the turnpike on their way to join General Pope when they encountered artillery fire from the woods to their left. An artillery duel ensued followed by a movement of Federal troops into the woods to their left. They had no idea they were facing Jackson's old brigade who had established a reputation with the north as one of the hardest fighting units in the southern army. As night fell the two sides fought to a bitter standstill with heavy losses on both sides. The various brigade commanders advised King not to stay the night—with the thought of continuing the battle the next morning—but to retreat toward Manassas where they hoped to join some of the other Federal divisions. He sent several messages to the other Federal commanders that he had found Jackson. Jackson spent the night disappointed that he didn't take greater advantage of the situation.

That same afternoon Lee and Longstreet reached Thoroughfare Gap, about ten miles west of Gainesville and only a day's march from Jackson's position. The gap was formed by large hills to the north and south. It was narrow enough that one or two regiments could defend it against a much larger force. Additionally, Chapman's Mill, a six story stone stacked structure, stood in the heart of the gap and served as a major vantage point for Federal sharpshooters.[4] Rickett's division was rushed from Haymarket to the gap with the mission of driving Longstreet back through the gap. If the Federals succeeded in their mission, it would take days for Longstreet to go further north and find another route. By then, Jackson would be overrun and the war might have ended.

Pope's first mistake was losing sight of Jackson and allowing Jackson to pick the ground for battle. His second was not paying attention to defending the gap. He knew Longstreet was coming and should have known that he would use the same route as Jackson. It

---

4. Chapman's Mill will originally built in the middle 1700's by slaves. Several stories were added to the structure prior to the Civil War using slave labor and the same stacked stone approach. Most of the mill still exists beside Interstate 66. It is the highest stacked stone building in the United States.

didn't take long for Longstreet to outmaneuver Ricketts and drive him back toward Gainesville. Knowing that it would be difficult to fight a head-to-head battle in the gap, Longstreet sent men over the top of the rugged, rocky hills north and south of the gap. Some of the force consisted of skirmishers from the 4th Texas. Ricketts soon recognized that he was outflanked. There was little choice but to return to Haymarket and await further orders.

Two comical events took place in the 4th Texas just before and after they passed through the gap. On August 26 and 27, the 4th marched for 30 miles from Waterloo Bridge through Salem to the approaches to Thoroughfare Gap. Completely exhausted the men camped for the night on a slope on the west side of the Gap. The men were dead tired and sound asleep when one of the officers at the top of the hill accidentally kicked a large empty barrel that had been filled with oats for the horses. The barrel went crashing down the hill toward the sleeping Texans. Frightened by the rolling barrel, an old gray mare loaded with kettles, tin cups and frying pans dashed down the hillside adding to the racket. The mare charged down the hill toward the Texans scattering pots and pans in all directions. The men awoke suddenly from their deep sleep and assumed that the enemy was in their midst. They deserted their guns and flags and tumbled down the hill toward a fence. After tearing down the fence and running in all directions, they finally stopped long enough to realize the reality of what took place. They returned to camp "convulsed with laughter."[5]

The incident led to the famous song "The Old Gray Mare Came Tearing Out of the Wilderness." It became the favorite marching song for the 4th Texas for the rest of the war. Over the years the title has been shortened to "The Old Gray Mare." The original words were written by Private Hames Augustus Bailey of the 5th Texas Regiment.[6]

---

5. John Bell Hood, *Advance and Retreat* (New Orleans: P. G. T. Beauregard, 1880), p. 38.
6. Harold B. Simpson, *Gaines Mill to Appomattox: Waco and McLennan County in Hood's Texas Brigade* (Waco: Texian Press, 1963), p. 96.

The second incident took place the very next night. J. B. Polley recounts that the Gap was occupied by the enemy when –

> . . . in the afternoon we approached it, and Law's brigade and skirmishers from Texas took an active part in the fighting that followed, and which resulted in the retreat of the enemy. The way clear, the Texas Brigade marched through the gap, following the railroad track, and shortly after sunset went into bivouac on a hill-side just beyond.
>
> An hour later, everybody except Bill Calhoun, of Company B, 4th Texas, was resting comfortably. He was an oddity of whom the Texas Brigade was proud, for although usually sad of countenance and melancholy of mien, in his bosom dwelt a spirit of drollery that was constantly effervescing and running over. His mess-mate and bed-fellow was Davidge. Carrying out a well-conceived plan for an equitable distribution of baggage, Davidge, on the morning of the 28th, was instructed with the transportation of the blankets and tent-cloths of the mess—Bill Calhoun with that of the provisions and the frying-pan. Davidge straggled, and when camping time came, was not on hand. Confident that he would soon put in his appearance, Bill prepared supper for the mess, and Davidge still remaining absent, ate it all himself. Then lighting his pipe, between puffs he chatted with such of his company as would listen and respond. The responses, after a while, growing few and sleepy, he declined an invitation of a friend to share the friend's blankets, and remarking that Davidge would surely be along soon, stretched himself out on the bare ground, and was soon asleep. But the night was cool enough to make some covering necessary and though Bill endured the hardness of his couch and the chilliness of the air without a murmur until midnight, he could endure it no longer. Standing erect in the midst of the 2500 recumbent forms that darkened the moonlit hill-side, he broke into magniloquent apostrophe:
>
> "Oh, Davidge, Davidge!" he cried, "friend of my bosom and possessor of my blanket, where art thou, Davidge, this cold and comfortless night? Art thou, indeed, false to thy many professions, false to their sacred obligations of the true and loyal friendship thou hast so often and fervently declared, and oblivious of duty, forgetful of the friend who has confided to

> thee even the well-worn blanket on which he dependeth for protection from the chilling blasts of winter? Art thou now peacefully and blissfully, but alas, ungratefully, reclining on some hospitable feather bed and dreaming of the joys that will be thine when this cruel war is over, or art thou, beguiled and betrayed by the demon of intemperance that hath bestowed upon thee such a damnable thirst for apple-jack, wallowing like a filthy and disreputable hog in the dirt before the door of some far-away mountain still-house, while I—thy friend and mess mate, thy boon companion in happiness and adversity—stand here alone, a homeless, houseless, blanketless orphan, his wandering and faltering footsteps guided only by the pale light of yonder refulgent orb of night, his shivering body covered only by the blue canopy of the sky, his restless slumber watched over only by the myriads of twinkling stars that shine in the heavens above him? Alas, Davidge, though hitherto trusted friend and companion and confidant of my youth and my manhood! Thou hast been weighed in the balance and found wanting. The surrounding and circumambient circumstances and facts furnish proof strong as holy writ, that I have been duped, deceived and outwitted, and ungratefully left to encounter the slings and arrows of misfortune alone and unsustained by any human aid." And dropping suddenly from the sublime to the ridiculous, Bill nudged the nearest man with his foot, and in a voice of entreaty that would have melted the hardest of hearts, said: "Say, Bill Hamby, roll over just a little bit, and let me get under the shadow of your blanket. If you don't ere the morning's dawn illumines the eastern horizon, I'll be a standing monument to man's inhumanity to man."[7]

Among the mirth there was serious business ahead. The noise of artillery fire and the smoke of battle could be seen and heard miles ahead. Jackson's corps was facing increasing odds and an ever-increasing series of frontal attacks. Fortunately these attacks were piecemeal, not well supported, but nonetheless fierce and deadly. Jackson's ability to hold his line was down to hours, perhaps a day at most. Ammunition was running low and his line was almost breached on several occasions. The unfinished railroad depression he had chosen for his

7. Polley, *op. cit.*, pp. 77-78.

defense was his saving grace. He needed Longstreet and the fighting Texans and he needed them soon.

Hood's brigade was chosen to lead the march to join Jackson. Hood placed Colonel Upton of the 5th Texas in command of 150 Texas riflemen to act as the advance guard. Upton moved quickly east on what is today highway 55 until he reached the Warrenton Turnpike (today Highway 29) at Gainesville. There he turned left on the Turnpike and marched the four miles to meet Jackson's right flank located just to the left of the Turnpike. This completed a tactical surprise and an unconventional dividing of units but a brilliant move on Lee's part to join his two forces.

Hood's two brigades were formed on either side of the turnpike, Law's brigade on the left and Hood's old brigade on the right. It took the remainder of that day (August 29) for the rest of Longstreet's corps to move through the Gap and arrive on the field of battle. As the units arrived they were formed to the right of the Texas regiments. The two corps formed a very large "V" with Jackson along the left half of the "V" and Longstreet along the right half. The Texas regiments were right in the middle, a position they often occupied in Lee's organization of the battlefield. The entire Confederate line, south to north, extended as much as five miles.

Throughout the 29th the federals continued to assault Jackson's line unaware that Longstreet was moving into position. Hood's men were mainly spectators until early that evening when Lee ordered Hood's Division and Evans' Brigade to attack the Federal left. Just prior to launching his attack, Hood was attacked by a group of Federals to his front. Hood counterattacked while "driving the foe through field and forest from position after position, till long after night had closed in upon the scene of conflict."[8] The advance of the Texas regiments was so fast and furious that they found themselves a mile into the Federal lines before darkness fell. This would normally be good news except for the fact that in the darkness, complicated by the woods, it was impossible to tell friend from foe. Hood's men were

8. Hood, *op. cit.*, pp. 34-35.

almost completely surrounded by the fires of Pope's army. Officers found themselves giving orders to troops from the other side.

J. B. Polley describes their predicament:

> At the word (halt), the brigade came to a sudden stop, the men standing motionless with wonder. A minute later the strange caution came whispered from man to man, all along the line, "Silence! We are surrounded by the enemy." It was the truth, and for a minute or more, a sadly humiliating truth, since to be surrounded presaged speedy capture and resulting shame and mortification. But our humiliation lasted but a second or two; with arms still in our hands, we could fight our way out, or die; and thus resolving, we asked each other in whispers how in the mischief we had got ourselves into such a trap. It had been easy to do so; moving forward on converging lines, the two brigades of the division had simply driven themselves, wedge-like, into the unoccupied space between two Federal brigades—the darkness of the night and the suddenness of the movement having prevented them from it, they moved closer to each other and thus closed the gap through which we had entered. But in doing that, they made the gap between themselves and the brigade on their left wider than was safe, and through this, after midnight, we stole, with bated breaths and noiseless steps, back to the line from which we had started.[9]

The losses to Texas brigade on the 29th were light. Colonel Work of the 5th Texas was among the wounded. In the darkness of the night, the colonel went forward to prove to his men that the dark figures to their front were Federals. It led to hand-to-hand combat during which the colonel was clubbed over the head with a rifle. He was back in command the next day. Unfortunately for the 4th Texas orator on the previous evening, Private Bill Calhoun was captured as a result of wanting to get a good look at them Yankees.[10]

Pope made another fatal error on the morning of the 30th. It was reported that wagons were moving along the Warrenton Turnpike toward Gainesville and away from the battle and that some of the

---

9. Polley, *op. cit.*, pp. 80-81.
10. Ibid., p. 82.

Confederates were moving in the same direction (probably Hood's men moving back to their lines very early that morning). In addition, Pope continued to dismiss reports that Longstreet was on the field in strength. Pope came to the conclusion that General Lee was in retreat and reported so to Washington. Unfortunately for Pope the opposite was true. Pope had recently been reinforced by Heintzelman's and Porter's corps from McClellan's Army. It seemed that everything was moving in his favor. Those reinforcements were countered by Longstreet's arrival. In spite of recent reports from the battlefield, Pope still had trouble believing that Longstreet was already there and that Pope's left flank was in serious danger.

Throughout the morning Pope renewed his attacks on Jackson's line, committing much of his reserve in a major effort to overwhelm Jackson. This created an opportunity for Longstreet to spring his attack in the hope that his units could overwhelm Pope's poorly defended left flank. Already the Confederate units were causing serious damage to the Federal lines. Hood was given the position of honor, so to speak. His regiments would fight just to the right of the Warrenton Turnpike with the 1st Texas beside the road, the 4th to their right, and the 5th to the right of the 4th. The rest of Longstreet's corps formed to the right of the Texans allowing them to pivot on the Texas brigade. The attack was not launched until 4:00 that afternoon. Hood recalled that the attack was "the most beautiful battle scene I have ever witnessed."[11]

Simpson gives an excellent summary of the fighting that day:

> After receiving orders to move forward, Hood started his division east along the Warrenton Pike—Law's Brigade advancing north of the pike and the Texas Brigade south of the road. The latter, in brigade line with the 1st Texas as the regiment of direction, moved rapidly across a wheat field and immediately engaged the skirmishers of G. K. Warren's Brigade of George Sykes' Division. The Texas Brigade, ordered to fire and then charge, quickly drove the skirmishers and their parent regiment, the Tenth New York Infantry, through

11. Hood, *op. cit.*, p. 37.

a rather heavy stand of timber and then across a field beyond the wooded area.[12]

In advancing through the timber the Brigade lost its regimental alignment, and the remainder of the battle was fought primarily by isolated regiments or by groups of regiments. Although the regiments remained generally within supporting distance of one another, each appeared to try to outdo the other in the rapidity of its advance and in the winning of battle honors. Both the 1st and 4th Texas fought most of the battle as isolated commands, while the Eighteenth Georgia, Hampton's Legion, and the 5th Texas fought as a unit through much of the engagement. After the Texans were relieved by Shanks Evans' Brigade on Chinn House Hill, however, the 5th Texas continued fighting as a part of Evans' command for a short time, then proceeded to fight as an individual unit until dark.

. . . The 4th Texas, under Colonel B. F. Carter [John Tatum's former Company Commander now elevated to Regimental Commander], continued to advance straight ahead upon emerging from the woods which most of the Brigade had entered in its pursuit of the Tenth New York. The Eighteenth Georgia, Hampton's Legion, and the 5th Texas, on the other hand, bore to the right after leaving the timber, scattering the Tenth and engaging the 5th New York Zouaves. Separated from the other regiments of the Texas Brigade, Carter's command turned its primary attention to "a battery stationed on the hill beyond the small creek [Young's Branch], supported by infantry in strong force who opened fire on [them]." Advancing at double-quick, the regiment crossed the stream, then reformed before charging the Pennsylvania battery commanded by Captain Mark Kerns. Although "greeted with a terrific fire of grape, canister, and musketry," and suffering heavy losses, the regiment routed the supporting infantry and killed or scattered the gunners. Many of the latter had continued to work their artillery pieces until the Texans approached "within forty paces."

Two outstanding acts of heroism took place during the charge of the 4th Texas on the guns of Kern's Battery. The attack on the Federal battery was led by Major William P. Townsend, original captain of Company C of the regiment. Regardless

12. Polley, *op. cit.*, p. 84.

of the deadly canister being fired at short range, Townsend, firing his six-shooter at the gunners, urged his men on until within a short distance of the guns his left foot was mangled by a direct hit. The wound necessitated the amputation of the lower leg and Townsend's subsequent return to Robertson County, Texas.[13]

Captain Mark Kerns died a hero's death. Although badly wounded and all of his guns deserted, Kerns continued to load and fire one of the artillery pieces himself until he was shot down. The Texans offered to take him to a field hospital for surgical attention but he refused to leave his guns saying, "I promised to drive you back, or die under my guns, and I have kept my word." Nevertheless, the badly bleeding captain was carried away from the scene of carnage by the Texans and laid under a large oak tree near Young's Branch. Colonel Carter, noting that the captain had little time to live, "left his overcoat for him to be buried in." Kerns died the next day and was buried under the large oak Tree.

After overrunning and capturing the guns that had eluded them at Gaines' Mill, the 4th Texas moved quickly to the valley beyond the hill where Kerns' Battery had been posted. Here they were forced to take shelter from the "sharp fire of muskety" directed at them from a nearby hill where Kerns' infantry support and surviving gunners had taken refuge. At the same time a Federal force moved upon Carter's left and rear compelling the Texans to retreat to a ravine where "for some time [they] maintained a steady fire on the enemy." However, the position of the 4th Texas soon became untenable as the enemy continued to build up on its flank and rear. In this situation, with "no prospects of supports," Carter withdrew his men under cover of the ravine and a hedge row back to Young's Branch, suffering several casualties in the process. The 4th Texas reformed under cover of the creek band and was at this time joined by the 1st Texas. Except for sending out a line of skirmishers to protect the captured artillery pieces of Kern's Battery, both Texas regiments remained near Young's Branch in battle position until relieved by General Evans' Brigade.[14]

13. Davis, *op. cit.,* p. 117.

14. Harold B. Simpson, *Hoods Texas Brigade: Lee's Grenadier Guard* (Dallas: Alcor Publishing Company, 1983), 148-150. Simpson occasionally quotes

There is an irony in the story of Captain Kerns that Simpson mentions briefly in his footnotes. It was Captain Kerns and his battery of artillery that faced the 4th Texas at Gaines Mill. When the 4th broke through the Federal line and raced up the hill they took some of the artillery pieces that belonged to Kerns. The captain barely escaped with his life and four of his artillery pieces. It was then that he vowed he would die at his guns rather than lose them to the Confederates. A few months later, Kerns was facing the same infantry regiment, the 4th Texas. They managed to capture the remaining four pieces that Kerns had saved earlier. The men of the 4th knew almost immediately who opposed them that day. If it had not been for the heat of the battle, they would have spared his life. They respected his heroism at Gaines Mill and at Second Manassas. Private Polley reports that "the watch, the keep-sakes and the letter he wrote were a few days later sent through the lines to the parties he named."[15]

The 5th Texas, with the help of the other regiments including the 4th, drove the Tenth New York ahead of them through the woods. The 5th New York Zouaves were in line behind the Tenth in a clearing on the south side of Young's Branch. The Texans had dealt with this unit before at Gaines Mill and with shouts across the river during the winter of 1861-62. They had both bragged that when they met each other in battle again that there would be no quarter given. As the Tenth New York retreated through the ranks of the 5th New York Zouaves, the 5th Texas suddenly appeared on the scene. The 5th New York only had time to get off one volley and it was generally over the heads of the Texans. The fire from the Texans was much more deadly; half the New York regiment was killed or wounded during the first volley. Their remaining numbers retreated through the creek and up the steep bank on the other side. Many of the remaining number were cut down as they waded across Young's Branch and scrambled up the opposite side.

John Hennessy in his remarkable book about the second battle of Manassas describes the aftermath following the attack:

from the official record (O.R.A.) Series I, Vol. 12, Part II, p. 616.

15. Polley, *op. cit.*, p. 90.

In the 5th New York of Warren's brigade, those who survived did not stop to regroup until they reached Henry Hill, a mile to the east. There Warren ordered Sergeant Chambers to jam the regimental flag into the ground. Slowly, sadly, the remnants of the regiment gathered around it. The dazed and shaken survivors presented a stunning sight. One of the Regulars who saw them remembered the scene particularly: "Warren sat immobile on his horse, looking back at the battle as if paralyzed, while his handful of men, formed in files of four, blackened with dust and smoke, stood under the colors as silent as statues, gazing vacantly . . . . A murmur of surprise and horror passed through the ranks of our Regulars at the fate of this brave regiment.

Forty minutes before, the 5th New York had counted more than five hundred men under its flag. Now it mustered just sixty. Some of the regiment would yet return, but most of them were dead, dying and wounded, scattered in the blood stained stubblefield a mile to the west. The scene of slaughter was "a ghastly, horrifying spectacle," said one of Hood's men. Another claimed that the Zouaves lay so thick that he could have walked from the top of the ridge to Young's Branch touching only the corpses of dead soldiers. Still another looking from afar, described the scene as striking, yet deceptive. The red breeches and blue jackets dotting the broom sedge reminded him of a Texas hillside covered with wildflowers in springtime.

During those ten awful minutes atop that ridge, the 5th New York lost nearly three hundred men shot—120 mortally. For a single infantry regiment it was the largest loss of life in any single battle of the entire Civil War. And it all happened in ten minutes. It was small wonder then that when the regiment's survivors gathered on the field forty-five years later to dedicate their monument, one of them said of that day: "War has been designated as Hell, and I can assure you that where the Regiment stood that day was the very vortex of Hell.[16]

[16] John J. Hennessy, *Return to Bull Run: The Campaign and Battle of Second Manassas* (New York: Simon & Schuster, 1993), p. 372. The Zouaves dressed in red pantaloons with a blue jacket and a fez for a cap. (The New York monument still stands on the spot where all this took place. It is interesting to note that the monument says that the New York regiment faced

Most of Hood's regiments moved forward without much resistance from the ridge that was occupied by Kern's battery to a ravine that opened up before Chinn Ridge. While they rested there a moment, the 4th spotted Hardin's Pennsylvanians huddled in a stand of pines on the north shoulder of the ridge. The 4th charged in their direction much like they did on Kern's battery. The Pennsylvania boys had enough fighting for one day and fled in the opposite direction.

Hood soon decided that the ravine did not provide enough cover. His position was still open to Federal cannons on Dogan's Ridge and the woods to his right were an invitation to other Federal forces to outflank his regiments. Hood made a quick decision to move his units to occupy the woods but in the haste of the moment the 4th did not receive the new orders. The 4th found itself in a difficult, lonely position where the battery on Dogan's Ridge could reign some deadly fire in their direction.

> Lieutenant Colonel B. F. Carter feared the Yankees might attack his left and rear, and urgently sent to Colonel Wofford of the 18th Georgia for help. Wofford replied that he could not come, that he was moving to the right. Rather than risk his command by holding his position alone, Colonel Carter marched his regiment back up over the knoll, past Kern's captured battery, to the cover of Young's Branch. There the 4th was joined by the heretofore wayward 1st Texas. And there both regiments would sit out the rest of the battle.[17]

Longstreet was doing his best to silence the artillery on Dogan's Ridge but he didn't have any practical way to make a direct charge. There were too few of his units on the north side of the turnpike. The alternative was to distract them with his own artillery located just south of the turnpike in the area of Groveton. Captain Reilly[18], always looking for a way to support as far forward as possible, moved

---

a Confederate division when in truth the main force they faced was the 5th Texas Infantry. Of course, the 5th Texas is not mentioned on the monument. The monument is located on the battlefield just south of Highway 29 and west of Highway 234.)

17. Hennessy, *op. cit.*, p. 379.

18. Reilly painting by Dale Gallon, available on his web site.

Reilley's Battery-Dale Gallon
Image courtesy of Gallon Historical Art: www.gallon.com

his pieces to the same knoll occupied previously by Kern's battery. From there he was able to accurately throw shells in the midst of the Yankee guns. In the midst of this shelling, one of Hood's staff arrived with an order. Reilly pointed toward Dogan Ridge and yelled to the officer, "See major, see! The domned skillipins skedaddle extensively—extensively sir!"[19] Reilly became famous on both sides of the line for his audacity, ability to be in the right place at the right time, and accuracy. Artillerymen to this day hold Captain Reilly in high esteem.

Hood was unable to control the advance of the 5th Texas as they joined the attack on Chinn Ridge with other units from Longstreet's army. They advanced as far as Sudley Road where it runs along the western base of Henry Hill. It was on Henry Hill that General Pope was making a last minute effort to save his army. He finally came to the conclusion that Longstreet was there in force, that Longstreet's

19. Mary Lasswell ed., *Rags and Hope: The Memoirs of Val C. Giles, Four Years with Hood's Brigade, Fourth Texas Infantry, 1861-1865* (New York: Coward-McCann, Inc., 1961), p. 131.

army was rolling up the Federal left flank, and if he, Pope, didn't do something quickly he could lose his whole army. He had no recourse but to order a general retreat back to Centreville. Several Federal regiments were ordered to cover the retreat from Henry Hill and the vicinity of Matthew's Hill, just to the north of Henry Hill. Between these two hills was the Warrenton Turnpike where it ran east from the main area of the battle over the stone bridge at Bull Run and on to Centreville, the same route that Highway 29 takes today from Warrenton, through Gainesville, through the battlefield, and on to Centreville. Late in the afternoon and into the evening the Yankee army marched in silence down this road to their secure defensive positions in Centreville.

Pope managed to extricate his army but his forces suffered dearly for his poor decision-making. According to Chaplain Davis "the loss of the enemy is astonishingly greater—thirty thousand at least in killed, wounded and prisoners."[20] While this figure may be somewhat exaggerated, Pope's losses were horrendous. Jackson had lost nearly four thousand men, Longstreet more than 4,700. These losses represented about fifteen percent of their force.[21]

Colonel Simpson outlines the heavy losses of the Texas Brigade:

> On the day following the battle, August 31, the Texas Brigade moved its bivouac to near Henry House Hill, cooked its rations, buried its dead, and took stock of its condition. The number of casualties had been great—628 killed, wounded, and missing. This would be the greatest battle loss that the Brigade would suffer during the war. Of the five regiments of the Brigade, the 5th Texas, which had engaged in the most sustained fighting on the 30th, suffered by far the heaviest casualties; suffering more, in fact, than any other regiment in Lee's Army—15 killed, 245 wounded and 1 missing. Seven color-bearers of Colonel Robertson's command had been killed or wounded, three companies of the regiment had fought the last phase of the battle without a commissioned officer present, and all three of the regimental officers had been hit. Colonel

20. Davis, *op. cit.*, p. 122.
21. Hennessy, *op. cit.*, p. 441.

> Robertson and Major King Bryan were wounded and Lieutenant Colonel J. C. Upton was killed. To show for this tremendous amount of blood-letting, the regiment could claim the capture of three stands of colors, two artillery batteries, and a large number of prisoners. The Bloody 5th had also left scores of dead and wounded Federals in its wake as it drove across the Manassas plains.
>
> Of the other regiments in the Texas Brigade, the Eighteenth Georgia, commanded by William T. Wofford, had the greatest loss—19 killed, and 114 wounded. The Georgians, who had played a major role in scattering the two New York regiments, lost three color-bearers on August 30. The 4th Texas, under Lieutenant Colonel B. F. Carter, reported 22 men killed and 77 wounded, while Hampton's Legion, led by Lieutenant Colonel M. W. Gary, had 11 killed and 63 wounded. . . . Colonel Work's 1st Texas Infantry, which except for a short period of time, was exposed only to artillery fire, listed only 28 casualties—10 killed and 18 wounded.[22]

According to the records kept by Chaplain Davis, John Tatum was with B Company, 4th Texas Infantry, throughout the battle of Second Manassas.[23] There is no indication in his military records that he was wounded. By the time the battle had ended, the men of the 4th looked more like beggars. Many were without shoes, trousers were torn, hair stuck through holes in their hats, and essential equipment was often missing. Some men became so eager to find shoes that they "shopped" for them among the Yankee dead on the battlefield. This was a common practice on both sides but for the southerners it was almost a necessity due to the lack of supplies from Texas and other states.

Not long after the battle Pope was removed from the command of the Army. His poor leadership had led to a bitter defeat, the second defeat for the Federals on the fields outside of Manassas. Pope was sent to a remote assignment in Minnesota for the remainder of the war. He spent his time trying to shift the blame to Porter but it was

22. Simpson, *op. cit.*, pp. 156-157.
23. Davis, *op. cit.*, p. 203.

clear to almost everyone that the major blame had to go to General Pope. McClellan's lack of support, or some would say "tepid support," didn't help the cause. How was McClellan rewarded? Lincoln reluctantly put McClellan back in charge of the Army.

Texas Regimental belt buckle

Chapter Six

South Mountain and Antietam (Sharpsburg)

Lee saw his victory at Second Manassas as an opportunity to take his military campaign north into Maryland. Several possibilities drove this decision. Some of the countries in Europe may see his recent victories and his aggressive move north as the right time to come in on the side of the southern cause and give them some support. Second, the northern population may see the move into Maryland as a major threat to Washington, a sign of bad times to come, and a reason to pressure Washington into a peace treaty. And third, there were many southern sympathizers in Maryland that might help Lee and his army with provisions and provide them with intelligence concerning the Federal movements. While all of this was sound reasoning at the time, almost none of it worked out to Lee's favor.

The Texas Brigade, greatly reduced in size by the recent battles, marched north with Longstreet's corps on September 1. Regiments now looked like companies, brigades like regiments, and divisions like brigades. The southern army proceeded northeast along the Warrenton Pike as far as Centreville. From there they turned north, passing near Germantown, through Dranesville and Leesburg. On the morning of September 5 they arrived at White's Ford on the Potomac River. The next day, September 6, they crossed the river with the 4th Texas band playing "Maryland, My Maryland."[1] General Hood, however, was not at the head of his division. On August 30th he sent some of his Texas scouts out into the countryside near Manassas where they captured a number of new Federal ambulances, all of them with a fresh team of horses. Just after Hood started north with his units, General Shank Evans instructed Hood to turn over the ambulances to his Carolina soldiers. Hood refused to obey the orders even though Evans was senior in rank. Hood regarded his division as the rightful owners since they were the ones that risked their lives for this valuable possession. Evans placed Hood under arrest and Longstreet ordered Hood to Culpeper for trial. Lee, however, changed the order to allow Hood to ride behind his troops but not in command until the matter could be resolved.[2]

The brigade marched through Buckeystown, Maryland, and bivouacked three miles south of Frederick. They spent the next two days bathing in the Monocracy River and destroying bridges over the river. They marched through Frederick on the 10th and moved northeast on the Washington Pike (the Old National Road). The next day they passed through the Catoctin Mountains, through Turner's Gap (South Mountain) to Boonsboro and on to Hagerstown. They camped just five miles below the Pennsylvania line.[3]

It was while they were making these moves that Lee made another bold decision that was contrary to military thinking. He divided his forces into three parts. He sent Jackson with most of his corps to Harper's Ferry in order to rid that important junction of all Federal forces. He kept a force near South Mountain, to include Jeb Stuart's cavalry, to protect the supply wagons and screen the enemy. Longstreet's command was near Hagerstown where Lee made his temporary headquarters. All of this might have worked to Lee's favor if a

---

1. J. B. Polley, *Hood's Texas Brigade* (New York: Neale Publishing Co., 1910), p. 112.
2. John Bell Hood, *Advance and Retreat* (New Orleans: P. G. T. Beauregard, 1880), p. 38.
3. Harold B. Simpson, *Hoods Texas Brigade: Lee's Grenadier Guard* (Dallas: Alcor Publishing Company, 1983), p. 163.

Union soldier had not found Lee's Special Orders 191, dated September 9, giving details concerning the various movements of the Confederate army. The orders had been wrapped around three cigars and accidently dropped near Frederick by a Confederate major. McClellan, normally slow to move, saw this as his opportunity to destroy Lee one piece at a time.

Late in the day, Lee received word that McClellan was moving rapidly from Frederick toward South Mountain. Lee soon guessed the reason for McClellan's swift movements. His suspicions were confirmed when J. E. B. Stuart sent word on the 14th that the Federal commander had a copy of Special Orders 191. Lee immediately ordered Longstreet's units to block the passes in South Mountain. He knew that McClellan's army was considerably larger and he could not afford to let McClellan cut his army in two. Longstreet's forces reached South Mountain on the 14th and found Hill's units heavily engaged with the Federal forces.

Just prior to their arrival near the mountain pass Simpson records the following:

> Upon approaching South Mountain, Hood still technically under arrest for his dispute with Evans and riding in the rear of his division as ordered, could hear the Texas Brigade in front of him start to chant, "Give us Hood! Give us Hood!" The individual shouts soon turned into a swell of voices, for as the command wound its way through the foothills of the mountain the entire division took up the cry. Hood did not comprehend the meaning of this appeal until he arrived with the rear of the column at a spot where he saw General Lee and Colonel R. H. Chilton watching the troops file by. As Hood drew near to the commander and his chief of staff, who were standing by a fence near the pike, the latter approached the Texan and told him that General Lee wished to see him. Dismounting, Hood walked over to Lee. The Confederate leader greeted Hood, an old friend, warmly and stated that he did not wish to enter battle "with one his best officers under arrest" and agreed "to suspend Hood's arrest until the impending battle was decided." As Hood rode past his old brigade, to assume his position at the front of his division once again, he

> was greeted by prolonged cheers and shouts even though the command was in the presence of the enemy.[4]

The Texas Brigade was originally ordered to take their position to the left of the Washington Pike but was later moved to the right. The confederates along this part of the line were being driven back by the advancing Federals. After finding their way through the heavy underbrush, Hood placed his units between Fox's and Turner's Gap. The Federals had crested the hill and were fighting their way through the trees and brush down the western side. Hood, perhaps remembering the charge at Gaines Mill, ordered the men to fix bayonets followed by an order "to front and charge."[5] The Federals to their front, part of Burnside's organization, were caught by surprise and quickly retreated back over the crest of the hill. Soon after the charge, darkness fell on the hillside and it became impossible to do much more.

Lee, knowing he didn't have the forces to hold the Yankees back much longer, ordered his units to fall back to a range of small hills west of Antietam Creek and just north of the town of Sharpsburg. The Texas Brigade was given the privilege of being a major part of the rear guard as Longstreet's corps made their way west. The casualties in the brigade on South Mountain were light. The 4th Texas records show that they had two wounded and six missing.[6] Simpson discovered that the muster rolls for the 4th Texas show that George Creed of Company E, originally listed as missing, was known to have deserted to the enemy.[7]

In the middle of the afternoon on September 16 both of Hood's brigades took positions on the left flank of Lee's army. The Texas Brigade, commanded by Colonel Wofford, was placed on the left near the Hagerstown Pike, and Law's Brigade to the right, almost parallel to and a little east of the Pike. Hood expected an attack before the day came to an end. His expectations were accurate. About two hours before sunset the lead elements of Hooker's corps arrived at Hood's

4. *Ibid.*, pp. 165-166.
5. Hood, *op. cit.*, p. 41.
6. *Ibid.*, p. 69.
7. Simpson, *op. cit.*, p. 168.

front. The artillery on both sides dueled for the better part of an hour before the infantry became engaged.

Colonel Wofford formed his brigade in line of battle with the 1st Texas in the center; Eighteenth Georgia, left center; 4th Texas, right center; 5th Texas on the right flank and Hampton's Legion on the left flank. He then ordered his men forward through a "brisk fire of shot and shell" across Miller's Cornfield to the East Woods. Here a "spirited action" took place between the Federals and the Brigade skirmish line supported by the 5th Texas under Captain Ike Turner. The Texans succeeded in driving part of the First Pennsylvania Rifles (Bucktails) and the Third Regiment of the Pennsylvania Reserves into the East Woods upon General Truman Seymour's Brigade, which was in support of the Federal skirmish line. With the coming of darkness the firing gradually subsided, and both sides bivouacked for the night on the battlefield—close enough for Hood's men to hear the enemy reinforcing his position in their immediate front. Hood's Division was the only Confederate force engaged on September 16.[8]

Hood's men, including the 4th, had eaten very little in the last three days, in most cases a little green corn. The general rode to Lee's headquarters and asked that two brigades be sent in his support so that he could feed his men and give them some rest. Hood was sent to General Jackson with his request. Some of Jackson's units had recently arrived from a successful skirmish at Harper's Ferry. He found Jackson asleep by the base of a tree. Jackson responded by sending Lawton's and Trimble's Brigades to Hood's relief. About 10:00 P.M. on the night of the 16th the Texas Brigade moved across the Hagerstown Pike to the West Woods behind the Dunkard Church in the hope of grabbing a substantial meal.[9]

The supply wagons were late in arriving with the needed provisions. It wasn't until very early the next morning that food was being prepared over the fires. Before they could all be fed, Hooker unleashed three of his divisions on the Confederate left flank. The

8. *Ibid.*, pp. 169-170.

9. Hood, *op. cit.*, p. 42.

Federal units with strong artillery support poured out of the North Woods and into the cornfield. Lawton's men, greatly outnumbered, were driven back in the direction of the church. Lawton was forced to send for Hood's division. Hood moved his men down the Pike and into the field across from the Dunkard Church. When they reached the battlefield "he was appalled to find that Lawton's command had melted away with the exception of General Harry Hays and some forty men who had rallied round him." Hood sent Hays to the rear so that he could pull together what was left of his command. Hood deployed Law's Brigade to the right and the Texas Brigade to the left. Hood's units extended for a quarter mile from the Hagerstown Pike on the left, across the cornfield, to the East Woods on the right.[10]

The details of the morning's battle are best told by those who were there. The first account comes from Lieutenant Colonel B. F. Carter, the commanding officer of the 4th Texas. His official report covers the activities of his regiment on the 17th.

> Soon after daylight the brigade formed line of battle in regular order, the 5th Texas being on my right and 1st Texas on my left, and about 7 AM were ordered to advance. I received no order as to which was the directing battalion, but, advancing diagonally to the right through the woods, we entered the open field on the right of the turnpike road. Here the fire upon us became severe, and, owing to our troops being in front of us, and the dense smoke pervading, we were unable to return the fire or see the enemy clearly. Still advancing, I came directly behind the Eleventh Mississippi, when I received the order from Captain Sellers for the Texas Brigade to halt. Halting, I ordered the men to lie down. At the same moment, the Eleventh Mississippi was ordered to advance, and a portion of two companies on my right, mistaking the order, advanced with them. After a moment I received an order from General Hood to move to the left until the left of my regiment rested on the crest, in advance, next to the turnpike road. Moving left-oblique in double-quick, I occupied the position indicat-

10. *Ibid.*, p. 43, and John P. Dyer, *The Gallant Hood* (Indianapolis: Bobbs-Merrill Co., 1950), p. 138.

> ed, and was then ordered by General Hood to move directly up the hill on the left of the troops then advancing.
>
> The enemy then occupied the hill in strong force, which receded before our steady advance. Arriving on the top of the hill, at the intersection of the cornfield with the turnpike, I found the enemy not only in heavy force in the cornfield in front, but occupying a ravine in the field on the left of the turnpike, from which position they poured a destructive fire upon us. I discovered at once that the position was untenable, but if I fell back the troops on my right who had entered the cornfield would be surrounded; so, wheeling my regiment to the left, I posted the men along the fence on either side of the turnpike, and replied as best we could to the tremendous fire of the enemy. We held this position for some time, until the troops in the cornfield on my right were falling back, when I ordered the regiment to move along the line of fence by the left flank. This movement, however, exposed us so much that we fell back directly under the hill. Here I ordered the regiment to halt and form, but at the same moment received an order from General Hood to move by the left flank into the woods. Forming here, I advanced on the left of the turnpike up to the fence at the edge of the field and rested in this position until I was ordered by Colonel Wofford to fall back to the point we started from in the morning, where the remnant of the brigade was formed. . .. .
>
> I cannot speak in too high terms of the conduct of both officers and men of my command. Exposed to a tremendous fire from superior numbers, in a position which it was apparent to all we could not hold, they fought on without flinching until the order to fall back was given. These men, too, were half-clad, many of them barefooted, and had been only half-fed for days before. The courage, constancy and patience of our men are beyond all praise.[11]

The second report comes from Captain Ike N. M. Turner who was in command of the 5th Texas during the engagements of the 16th and 17th.

> About 8 o'clock at night (on the 16th) we were relieved, and retired to the woods in rear of the church. Slept until about

[11.] Polley, *op. cit.*, pp. 121-123.

Texas Pride-Dale Gallon
The 4th and 5th Texas defending Lee's left flank near the cornfield
Image courtesy of Gallon Historical Art: www.gallon.com

day, when firing commenced in front. We were called to attention; thrown around the hill in line of battle to protect us from grape and shell. We had not occupied this position more than half an hour before we were ordered out as support for the Third Brigade. We caught up with said brigade where our first line had been fighting. Here the 5th was ordered to halt, by Major (Captain) Sellers, and allow the regiments on the right of the Third to advance. While lying here, General Hood rode up, ordering me to incline to the right, press forward, and drive the enemy out of the woods, which we did. The enemy twice tried to regain their position in the woods by advancing a force through the lower edge of the cornfield, which we repulsed. From a point of timber about 400 yards to our front and left, I discovered strong reinforcements marching out by the left flank down a hollow, which protected them from our fire. Allowing them to get within 75 yards of us with lines unbroken, I saw we would soon be hard pressed. Sent four times to Major (Captain) Sellers for support, determined to hold my position as long as possible. My men were out of ammunition, the enemy not more than 100 yards in my front, no support, no ammunition; all of our troops had fallen back on my left; I deemed it prudent to fall back also.

> Officers and men, with few exceptions, behaved well. The casualties of the regiment were 5 killed and 81 wounded.[12]

The brunt of the battle fell on the 1st Texas. It is clear from the following report that few units, if any, have faced the galling fire that fell on them on the 17th. You can almost hear the emotion in Lieutenant Colonel P. A. Work as he files his official report.

> The brigade, having been formed in order of battle upon the ground occupied by it on the night of the 16th, in the following order, to wit, 1st Texas in the center, Eighteenth Georgia left center, 4th Texas right center, 5th Texas on the right flank, and Hampton's Legion on the left flank, was moved forward to engage the enemy about -- o-'clock, the latter having made an attack upon our forces occupying a position in front of this brigade. Advancing through the woods some 200 yards, under a heavy fire of grape, canister and shell from the enemy's artillery, the brigade emerged into an open clover field some 200 to 250 yards in width, across which the forward movement was continued for some 150 to 200 yards, when, it being discovered that the left flank of the brigade was exposed to attack, I was ordered to move by the left flank, following a corresponding move of the Eighteenth Georgia and Hampton's Legion upon my left, which I did until ordered to move by the right flank, which was also done.
>
> Advancing now by the right flank (my original front), I entered a cornfield and soon became engaged with a force of the enemy, driving them before me to the farther side of the cornfield. As soon as the regiment became engaged with the enemy in the cornfield, it became impossible to restrain the men, and they rushed forward, pressing the enemy close until we had advanced a considerable distance ahead of both the right and left wings of the brigade. Discovering that this would probably be the case when my men first dashed forward, I dispatched you two different messengers, to wit--Capt. John R. Woodward, Company G, and Private A. G. Hanks, Company F-- stating that I was driving the enemy and requesting you to hurry up the regiments on my right and left to my support. It was not until we reached the farther side of the cornfield that I could check the regiment. By this time we had broken the first

12. *Ibid.*, p. 123.

line of battle of the enemy and had advanced to within some thirty steps of his second line, secreted behind a breast- work of fence rails thrown up in heaps upon the ground, when a battery of artillery some 150 or 200 yards in our front was opened upon us. My men continued firing, a portion of them at the enemy's men and others at the artillerists, the result of which was that the enemy's second line broke and fled, and the artillery was limbered up and started to the rear, when the whole fire of my regiment was concentrated on the artillerists and horses, knocking over men and horses with such effect that the artillery was abandoned. Very soon, however, a force of the enemy was moved up to the support of this artillery, when it again opened fire upon us.

Just at the farther side of the cornfield was the point where I was in great doubt as to the proper move to be made by me. I was aware that my regiment had advanced 150 or 200 yards farther than the regiment upon my left, so diverging as to leave a wide interval between the right flank of the Eighteenth Georgia and my left, thus exposing both regiments to attack--the Eighteenth upon the right and the 1st Texas upon the left flank. I was aware at the same time that a heavy force of the enemy was massed upon my left, and felt confident that in case I moved farther to the front I would be attacked upon my left and rear and annihilated. Had I moved forward to carry the enemy's battery I would have exposed the regiment to attack from three different directions, to wit, from the front from infantry and artillery and upon the left and rear from infantry. I am told also by some of the men that had I advanced a little farther to the front my right flank would have become exposed to attack, and am assured that some distance to my front, and obliquely to my right, was a large force of the enemy. This I did not discover myself. At this juncture I dispatched Acting Adjutant W. Shropshire to say to you that unless the regiments upon my left were moved up quickly to my relief and support upon my left, I would be forced to abandon my position and withdraw. Before the return of Shropshire a fire of musketry was opened upon me from my left and rear, which determined me at once to withdraw, as I had but a handful of men left, all of which must have been slain or captured had I remained longer. I at once gave the order to fall back and the few men remaining to me retired, turning to fire upon the enemy as rapidly as their pieces could be loaded and fired.

I entered the engagement with 226 men, officers (field and staff) included, of which number 170 are known to have been killed and wounded, besides twelve others who are missing, and, doubtless, also killed or wounded.

During the engagement I saw four bearers of our State colors shot down, to wit: First, John Hanson, Company L; second, James Day, Company M; third, Charles H. Kingsley, Company L, and fourth, James K. Malone, Company A. After the fall of these, still others raised the colors until four more bearers were shot down. Not having seen plainly who these were, I am unable to give their names in this report, but will do so as soon as, upon inquiry, I can ascertain.

It is a source of mortification to state that upon retiring from the engagement our colors were not brought off. I can but feel that some degree of odium must attach under the most favorable circumstances, and although such are the circumstances surrounding the conduct of this regiment, the loss of our flag will always remain a matter of sore and deep regret. In this connection it is but proper to state, in addition to that detailed in the above and foregoing report, the additional circumstances and causes which led to its loss. When the order to retire was given, the colors began the movement to the rear, when the color-bearer, after roving but a *few* paces, was shot down. Upon their fall some half dozen hastened to raise them, one of whom did raise them and move off, when he was shot down, which was not discovered by those surviving. While falling back, and when we had nearly reached the clover field hereinbefore alluded to (being still in the cornfield), I gave the order to halt, and inquired for the colors, intending to dress upon them, when I was told that the colors had gone out of the cornfield. Then I gave the order to move on out of the corn and form behind the crest of a small ridge just outside of the corn and in the clover field. It was when I reached this point that I became satisfied our colors were lost, for I looked in every direction and they were nowhere to be seen. It was then too late to recover them. There was no one who

> knew the spot where they had last fallen, and, owing to the density of the corn, a view of no object could be had but for a few feet. By this time, also, the enemy had moved up and was within some thirty-five or forty yards of my left (proper) and rear, and another force was following us. No blame, I feel, should attach to the men or officers, all of whom fought heroically and well. There was no such conduct upon their part as abandoning or deserting the colors. They fought bravely, and unflinchingly faced a terrible hail of bullets and artillery until ordered by me to retire. The colors started back with them and when they were lost no man knew save him who had fallen with them. It is perhaps due to myself to state that when I determined to retire I requested Captain (US) Connally to give the order upon the right and stepped to the left to direct Captain Woodward to give the order upon the left, from which point I moved on to the extreme left, to discover, if possible, the locality of the enemy attacking from that quarter, in order to be prepared to govern the movements of my regiment, so as to protect it as far as possible from danger and damage. While I was at the left thus engaged, the regiment commenced the movement to the rear, and, not being near the center, I was unable, owing to the density of the corn, to see where the colors were and when they fell.
>
> Capt. John R. Woodward of Company G acted in the capacity of major during the engagement, and aided me greatly in directing the movements of the regiment. Major (Matt) Dale, acting as lieutenant-colonel, had moved from the right and was conferring with me as to the propriety of advancing or at once withdrawing, when he was killed. Feeling that it was madness to advance with the few men left, I remained for several minutes after the fall of Major Dale, awaiting orders and information as to what my movements should be, being unwilling to withdraw as long as I had the ability to hold my then position without orders to do so.[13]

The final report comes from Private W. R. Hamby, a member of B Company, 4th Texas Infantry Regiment. Hamby fought alongside John Tatum. The two of them would have been part of the small number that remained in B Company after Second Manassas. Hamby's

13. *Ibid.*, pp. 124-128.

report of the battle, the same that would have been written by John, was published after the war.

> After the battle of South Mountain, September 14, we were the rear guard of the army on the march to Sharpsburg. On the morning of September 15, with a detail of one hundred men under Major Sellers, I was with the rear guard of the rear guard; and after the army crossed the Antietam, we were on the skirmish line along the west bank of that stream, until the 16th. In the meantime the brigade had formed a line of battle along the Hagerstown and Sharpsburg Turnpike, near the Dunkard church. This modest and hitherto unknown church was destined soon to become historical, as it was the storm center of the great battle fought September 17, 1862, called Sharpsburg by the Confederates, and Antietam by the Federals. The church was about a mile north of the town of Sharpsburg and about a mile west of the Antietam River. From the church north, along the west side of the pike, the woods extended about a quarter of a mile to an open field, extending still farther north several hundred yards. Across the pike east of the church were open fields, somewhat rocky and hilly, extending about half a mile north, and intersecting with a cornfield. East of the fields were woods extending toward the river.
>
> About sunset, the evening of the 16th, the Federal skirmish line was seen advancing through the woods east of us, closely followed by lines of battle in echelon with banners waving, drums beating, and bugles blowing. It was a magnificent spectacle, and looked more like they were on a grand review than going to battle. Our thin single line presented a striking contrast. Since leaving Richmond, about one month previous, we had marched over two hundred miles, and had participated in engagements at Freeman's Ford, Thoroughfare Gap, Second Manassas, and South Mountain, and had lost six hundred and thirty-eight men, killed and wounded. For the past several days, we had subsisted chiefly on apples and green corn. Many of us were barefooted and ragged, and all of us were foot-sore, weary, and hungry, but full of patriotic ardor and inspired faith in the justice of our cause.
>
> The fight was opened by the artillery on our right, between us and Law's Brigade, which was composed of the 4th Alabama,

6th North Carolina, 2nd Mississippi, and 11th Mississippi. They were as gallant soldiers, collectively or individually, as ever fought a battle. Among the first to enter the field, they were on the firing line when the last shot was fired. Both brigades advanced across the field with our skirmish line in front, which fell in with the main line as we entered the woods. The action continued for some time after dark; and when firing ceased, the two lines were so close together that they could hear each other speak. We knew this was only a preliminary skirmish, as we could tell from the sounds in front of us that the Federals were massing their troops for a desperate battle the following day. In this position we remained until far into the night, when we were relieved by General Lawton's division, and marched a short distance to the rear. After a long delay, some flour was issued to us, which was the first ration of any kind we had received since leaving Hagerstown; but before the flour could be cooked and eaten, the battle of Sharpsburg had begun.

It was scarcely daylight Wednesday morning, September 17, when the Texas Brigade was ordered in line of battle, and by sunrise it had crossed the pike in front of the Dunkard church and entered the meadow to take the place of the troops who had relieved us only a few hours before. The 5th Texas was on the right of the brigade, and as it entered the field was ordered into the woods east of the cornfield, where the fighting had occurred the previous evening. The 4th Texas, 1st Texas, 18th Georgia, and Hampton's Legion entered the meadow in the order named, and at once encountered a heavy fire. The troops in front had lost half their numbers, had exhausted their ammunition, and were retiring, and the smoke was so dense that the enemy could scarcely be seen to return his fire. The 4th Texas was ordered by the left flank; to the left of the brigade, up the side of a hill towards the pike. In this formation, the 4th Texas, Hampton's Legion, 18th Georgia, and 1st Texas advanced and drove the Union lines out of the open fields, back upon their reserves across the pike on the west and beyond the cornfield on the north.

The enemy's reinforcements appearing in strong numbers on the left, the 4th Texas changed from front to left flank and took position along the pike near the south edge of the cornfield. A short distance to the rear were some stone boulders, behind which some of our wounded were placed to protect them as far as possible from further injury; but even then several were struck the second and some the third time. Hampton's Legion and the 18th Georgia were farther into the cornfield, facing a galling fire from infantry and artillery with a steadiness unsurpassed. The 1st Texas had advanced some distance beyond the remainder of the brigade toward the north side of the cornfield, breaking two lines of the enemy and forcing them to abandon a battery and take shelter in the ravine north of the field. Three times the enemy tried to check the 5th Texas in the woods east of the cornfield, and each time broke and fled before their intrepid advance.

The Texas Brigade was now only a skirmish line; in fact, all of the Confederates on this portion of the field scarcely covered a fourth of the Federal front. It was yet early in the morning, although the battle had been hot and furious for some hours. In addition to the infantry and artillery on front and flanks, the heights above the Antietam were crowned with long-range batteries that poured a merciless fire; while the fresh troops of the Union forces seemed inexhaustible as they were thrown upon the fragments of the Confederate lines. The earth and sky seemed to be on fire, and it looked like here would be the Thermopylae of the Texas Brigade. With sublime courage the 1st Texas held their advanced position in the cornfield against overwhelming numbers, and retired only to escape annihilation. Unsupported and with both flanks uncovered, the 4th Texas, Hampton's Legion and the 18th Georgia met the advancing enemy from across the pike and drove them back and held their line. Many of the men had exhausted their ammunition and supplied themselves from the cartridge boxes of the dead and wounded around them. They were holding a position they knew they could not retain; yet men never fought better, and they withdrew only to keep from being surrounded. Falling back slowly below the crest of the hill, the line moved through the field, crossed the pike and took position in the woods near the church. The 4th Texas was then ordered up through the woods west of the pike near the edge of the field on the north, where they remained about an hour, defiantly waving their flag over empty muskets, when they were ordered to rejoin the other regiments of the brigade.

The 5th Texas, finding their ammunition exhausted and that they were being flanked, retired and also rejoined the brigade. By this time the morning was far gone, and the Federals had advanced down both sides of the pike within a short distance of the line held by the remnants of Hood's division, who stood facing them almost exhausted and practically without ammunition.

At last the long-looked-for reinforcements arrived, and again the enemy was driven back upon their reserves. The Texas Brigade was then ordered a short distance to the rear for a fresh supply of ammunition, and again returned to the front about noon and found the woods near the church lately occupied by them, in possession of the enemy; but as our line advanced, the Federals fell back across the pike into the field, about three hundred yards beyond the church. We steadily held our line near the pike until about sunset, when we were moved a short distance to the right, where we remained in line of battle until the night of the 18th, when the entire army withdrew and re-crossed the Potomac into Virginia.

If the reinforcements had reached the firing line before the Texas Brigade and Law's Brigade were forced to abandon their advanced positions, the Federals would have been swept from the field and another triumph would have been added to the list of Confederate victories. Our dead lay in rows upon the ground, where they had fought a fruitless fight; and instead of a Confederate victory, it was an indecisive contest, giving hope and courage to the Federals and depressing in its effect upon the Confederates.

The battle of Sharpsburg was fought with desperate courage by both the gray and the blue, and the 17th of September, 1863 [1862], stands out conspicuously as the bloodiest day in American history.... More men were killed and wounded that day than any other one day during the war between the States, and I doubt if the dead and wounded ever lay thicker upon any field than was seen from the old Dunkard church north, for more than half a mile. The action commenced about daybreak, and by sunset the bloody work had ended.

The 1st Texas went into battle with 226 men, and lost, in killed and wounded, 186, a loss of eighty-two per cent. As one flag-

Confederate dead on the fence line near Hagerstown Pike and near the cornfield. The dead may be members of the 4th Texas.
Library of Congress

> bearer would fall, another would seize the flag, until nine men had fallen beneath their colors. Official records show that the 1st Texas lost more men, killed and wounded, in the battle of Sharpsburg, in proportion to numbers engaged, than any other regiment engaged, either Federal or Confederate, in any other battle of the war. The 4th Texas went into the fight with 200 men, and lost 107; the 5th Texas went into the fight with 175 men, and lost 86; the Eighteenth Georgia went into the fight with 176 men, and lost 85; Hampton's Legion went into the fight with 77 men, and lost 55, including four flag-bearers. In the aggregate the Texas Brigade went into the fight with 854, rank and file, and lost 519 killed and wounded, including sixteen flag-bearers, a loss of over sixty per cent. This does not include the 'missing,' many of whom were, no doubt, killed or wounded.[14]

The Federal losses in the cornfield and along the Pike were extremely heavy. They soon realized that they were paying too high a price for this piece of land. As more Federal units arrived on the

---

14. *Ibid.*, pp. 129-134.

Dead Confederate artillery men near Dunkard Church
Reilly's battery was near this location but the men here are from a different unit.
Library of Congress

battlefield the attacks shifted from Lee's left flank further to the right. Fierce battles continued throughout the day. It can best be described by looking at the battlefield like the face of a clock: at 12 o'clock is the Hagerstown Turnpike and the cornfield, at 3 o'clock is the bloody lane, at 5 o'clock is the old stone bridge (Burnside's Bridge), and at 6 o'clock rests the outskirts of Sharpsburg and the Pike headed south through the town. The attacks on the 17th of September, 1862, started early in the morning at the turnpike and the cornfield and continued around to bloody lane. By late afternoon Federal forces were fighting their way across the stone bridge, and by evening the battle finally came to an end near the town.

That evening Lee's senior officers advised him to move south across the Potomac back into Virginia. The Potomac was a short distance behind them. Lee went against their advice and decided to remain on the battlefield throughout the 18th. If McClellan wanted another fight, Lee was willing to give it to him. McClellan, however, was satisfied with what he had accomplished and was not prepared for another battle. The day was spent with both sides treating their wounds, burying their dead, and reorganizing their forces.

Casualties were so heavy in the Battle of Antietam that September 17, 1862, is termed the bloodiest day of the Civil War. In fact, it was the bloodiest day in the history of the United States. Of McClellan's 26,023 killed, wounded, and captured during the Maryland Campaign (including Harpers Ferry), he counted 12,410 at Antietam. Of Lee's 13,385 casualties during the campaign, 10,700 fell at Antietam. The Texas Brigade went into battle with 854 men and lost 560 in killed, wounded and missing. This is a casualty rate of over sixty-four per cent, the third highest in the Civil War. When Hood reported to Lee the evening of the 17th he was asked about his "splendid division." Hood replied, "They are lying on the field where you sent them, sir; but few have straggled. My division has been almost wiped out!"

There is no indication in John Tatum's record that he was wounded although injuries that did not require hospitalization were not always noted. We know that he was on the battlefield at South Mountain and at Antietam on September 16 and 17.[15] He must have looked over the battlefield and mourned the loss of many of his friends. After the battle of Gaines Mill, Second Manassas, South Mountain, and Antietam, most of B Company were now gone. He had survived but for how much longer? Was he ever to return home?

15. Chaplain Nicholas A. Davis, *Chaplain Davis and Hood's Texas Brigade,* Donald E. Everett ed. (San Antonio: Principia Press of Trinity University, 1962), p. 203.

President Lincoln meeting with General McClellan a few days after the battle of Antietam
Library of Congress

Chapter Seven

The Last March

The 4th Texas and the Texas Brigade were in dire need of some rest and recuperation. Longstreet and Hood recognized the sad condition of their units and planned on giving them the needed rest as long as McClellan did not engage in a hot pursuit as they crossed back over the Potomac into Virginia. Fortunately McClellan obliged them by dragging his feet. He only pursued until he met some token resistance from some of Jackson's units. Longstreet took this opportunity to move his corps through what is now the panhandle of West Virginia to a few miles north of Winchester, Virginia, near a clear, running lake. It was there that the 4th Texas spent the better part of October resting from the intense campaign schedule that had started in June at Gaines Mill.

The month of October, 1862, was spent carrying out light camp duties, reorganizing units, and occasional drills. During this time some of the men wounded at Gaines Mill and Second Manassas returned to duty. A few of the lightly wounded at Antietam joined them as well. By October 8, Longstreet felt that his corps was fit enough to hold a formal review for the citizens of Winchester and the surrounding area. Many well known citizens were in attendance, to include General Lee and Hood. Twenty-one brigades consisting of eighty regiments passed by in review. Although the Texans passed by proudly, their appearance left a lot to be desired. New uniforms had been distributed to a few but many of the men were still in their ragged clothing and without shoes.[1]

A plea went out by way of a letter from Chaplain Davis to the citizens of Richmond pleading for shoes.

To the Editor of the Whig:

> I have just arrived from Fredericksburg: the prospects are good for a fight, but our men are not all shod. On yesterday evening an order was read on dress parade to the effect that, being barefooted would not excuse any man from duty. Those who were without shoes were ordered to make moccasins of raw hide, and stand in their places; and we feel that Texans will come as near discharging their duty as any who will meet the next struggle; but I ask the good people of Richmond and surrounding country, if they will stand by and see them go into the fight without shoes. We are too far from home to look to our friends there for help—We acknowledge the kindness shown us last winter, and many of the recipients have poured out their life's blood on the soil of Virginia.
>
> We are from the far South, and the cold is severe to us. It will require at least one hundred pairs of shoes, and five hundred pairs of socks to complete one suit for our men. Those who are disposed to contribute, will please send forward their mite to the depot of the Regiment, on 15th street, between Main and Cary, over Ratcliff's and it will be forwarded immediately.
>
> N. A. Davis
> Chaplain, 4th Texas.[2]

It was during this time near Winchester that the 4th Texas decided it was time to retire their regimental flag and replace it with a new one. It is impossible to overestimate the importance and devotion shown to the regimental flags throughout the Civil War. It was their rallying point in battle, their point of alignment, and their reminder of their cause. The 4th Texas regimental banner had sixty-five bullet

1. Harold B. Simpson, *Hood's Texas Brigade: Lee's grenadier Guard* (Dallas: Alcor Publishing Company, 1983), p. 187.

2. Chaplain Nicholas A. Davis, *Chaplain Davis and Hood's Texas Brigade,* Donald E. Everett ed. (San Antonio: Principia Press of Trinity University, 1962), p. 178

and three shell holes piercing its colors. Chaplain Davis reported that "nine flag bearers had fallen beneath its folds."[3] The flag was sent to the governor of Texas with this letter:

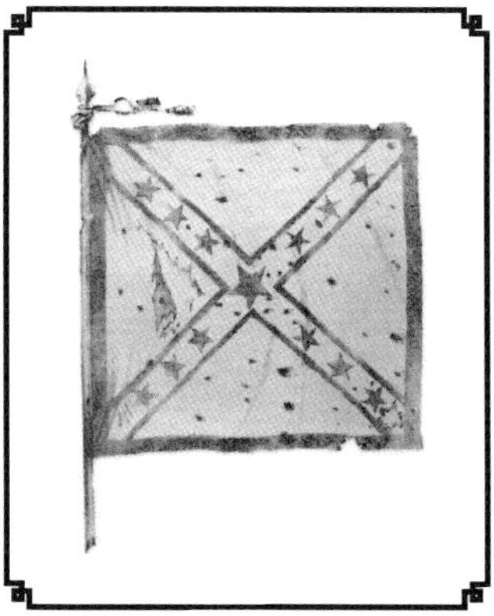

To His Excellency, F. R. Lubbock, governor of Texas.

SIR: I have the honor to present to you, by the hand of Captain S. H. Darden, the battle-flag of the 4th Texas Regiment, borne by them in the battles of Eltham's Landing, Seven Pines, Gaines' Farm, Malvern Hill, Freeman's Ford, Manassas Plains, Boonsboro Gap and Sharpsburg. From its torn and tattered condition, it can no longer be used; and it is returned to you, that it may be preserved among the archives of the State, as a testimonial of the gallantry of her sons, who have fought beneath its fold. I need not dwell upon the services of my Regiment. Its deeds in battle will go into the history of our country, and speak for themselves. And this silent witness bears eloquent evidence, that the men who followed it in action, were where shot fell thick, and death was in the air.

You will readily believe, Governor, that we part from the old flag with painful feelings. More than five hundred of our comrades in arms have fallen beneath its folds. And it is to us an emblem of constancy under multiplied hardships, gallant and dauntless courage in the storm of battle, and devotion unto death to our cause. Let it be preserved sacredly, that the remnant of our little band may, in future days, gaze upon its battle-

3. *Ibid.*, p. 135.

stained colors, recall to mind the sufferings they have endured in their country's cause, and their children incited to renewed vigilance, in the preservation of those liberties for which we are contending.

Our General has presented us with another "battle-flag," and we hope to be able to acquit ourselves as well with that as we have done with the old one.

Respectfully your serv't,

B. F. Carter

Lieut. Col. Commanding.[4]

Longstreet's Corps left the Winchester area on October 29. They marched southward toward Front Royal, through Manassas Gap, and on to Culpeper where they arrived on November 1. Two days later the Texas Brigade moved to the vicinity of the old battlefield at Cedar Mountain about six miles south of Culpeper. They remained in this position until mid- November. On November 7, word was received that General Ambrose Everett Burnside replaced General McClellan as the commander of the Federal Army of the Potomac. Unfortunately for President Lincoln, this choice was not an improvement. Burnside quickly moved his army toward Fredericksburg, Virginia, in the hope that he could place his units between Lee and the city of Richmond. Lee quickly discerned Burnside's plan and ordered Longstreet's corps to Fredericksburg. The Texas Brigade was the last unit to leave their position near Culpeper. They departed on November 19 and arrived near Fredericksburg on the 22nd.[5]

Unfortunately John Tatum did not depart with his unit. He was stricken with Febris Typhoides and sent to the hospital in Richmond on November 14, 1862. This acute disease was well known to physicians at the time and was the cause of death for thousands of Civil War soldiers. It was caused by close contact or by ingestion of food or water contaminated with feces (Salmonella typhi bacteria) from

4. *Ibid.*, pp. 136-137. Lieutenant Colonel, John Tatum's first company commander, died as a result of wounds he received near "Little Round Top" during the battle of Gettysburg, July 2, 1863.

5. Simpson, *op. cit.*, p. 190.

an infected person causing a high fever, headache, abdominal pain, and either constipation or diarrhea. The details of what caused this illness were guessed at by many military doctors but its exact cause was not well-known.

John was sent to the Alabama Hospital in Richmond, also known as Hospital No. 20. It was located at 25th and Franklin Streets and was also known as the Tobacco factory building of Turpin and Yarbrough. It was opened as a hospital in June 1862 and was still operating in late 1863. Like many of the hospitals quickly created in Civil War Richmond, it was originally a warehouse. It has since been torn down although the 2nd Alabama hospital still exists across the street. The building housing the 2nd Alabama hospital was built similar to Hospital No. 20 and had the same capacity. Normally John would have been sent to the Texas Hospital (Hospital No. 25) located in the center of the block south of Main Street, between Pear and Peach Streets. This hospital, like most in the city, would have been filled to capacity with wounded and sick from Second Manassas and Antietam. Chaplain Davis was the main driving force behind the building of the Texas Hospital. Records at Hospital No. 20 imply that John was admitted on October 31, 1862, but the 4th Texas record of No-

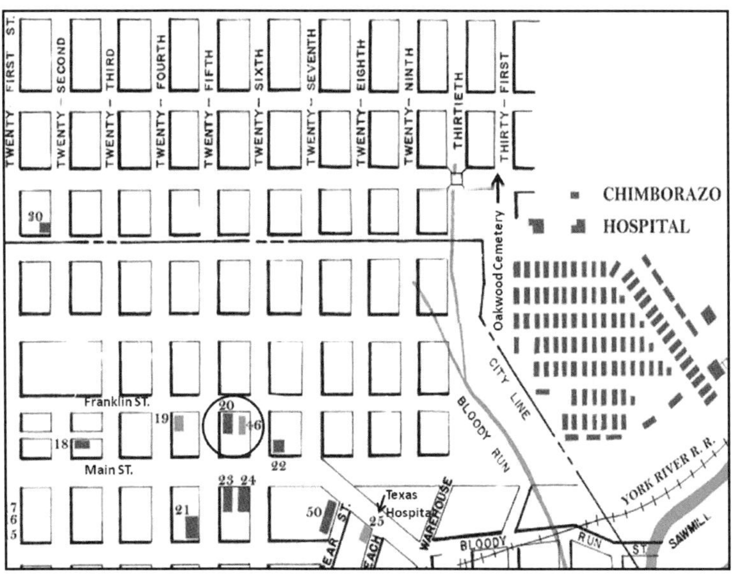

vember 14 is probably more accurate. John's military record indicates that he had $8.60 in Confederate money when he was admitted.[6]

The ladies of Richmond assisted at many of the hospitals but they were too few in number to meet the growing need. In many cases, including Hospital No. 20, slaves were used to care for the Confederate soldiers. John was in the hospital for a total of seven weeks and most of his care would have been entrusted to a slave.[7]

John's symptoms during his stay could have been taken out of the worst horror movie. Dr. Lilienthal, a prominent physician who lived during the Civil War, wrote a book that describes in detail the symptoms that doctors in his day observed as patients suffered with Febris Typhoides. His observations and research go on for many pages. Recorded below is his description of the first stage:

> Obstinate autumnal fever, with swollen spleen and liver; sleeplessness from congestion at the base of the brain; wants to walk about, seeing all sorts of visions. Acetic-acid [Acet-ac] fever with violent delirium, incoherent talking, with distended and obstinate constipation or with rumbling of bowels, colic and diarrhoea, followed by stupor, interrupted by delirious talking; slow, putrid fever with night-sweats. Intense thirst for water with passing large quantities of urine; vomiting of all nourishment with sensation as if there were a sore spot or ulcer in stomach; profuse, very weakening diarrhoea in later stages of fever; febrile heat with dry, hot skin and oppressed breathing during first stage; great debility and prostration. [8]

These symptoms continue to grow worse until the patient is unable to communicate, kidneys completely fail, and the patient slips into a coma. This would have been John Tatum's fate as his illness progressed. No longer able to fight the disease, John died on January 5, 1863. His body was moved to a pine box, placed on a wagon, and sent to Oakwood Cemetery for burial. Oakwood was straight down

6. See Appendix A.

7. Several months after John's death, a notice appeared in a Richmond newspaper asking slave owners to send more slaves to work in Hospital No. 20.

8. Dr. Samuel Lilienthal, 1815-1891, *Homeopathic Therapeutis,* p. 486.

Hospital 20 was torn down but would have looked almost exactly like this recent picture. The photo is the 2nd Alabama Hospital, a historical building still standing. It was located across 25th Street from Hospital 20. Both hospitals were former tobacco warehouses and built to look alike. The photo was rotated 180 degrees and placed on the corner that was occupied by No. 20.

21st Street to the edge of the city. It was there that the vast majority of Confederate soldiers who died in the Richmond hospitals were buried. The Richmond Dispatch, September 9, 1862, records that 4882 soldiers were buried at Oakwood Cemetery between September 1, 1861, and September 1, 1862. Many of them were placed in common, unmarked graves. Such was the case with John. No record has been found that show he was placed in a marked grave although some members of the 4th Texas Infantry Regiment did receive a marked headstone and mention in the graves registry maintained by the cemetery.[9]

Oakwood Cemetery was constantly plagued with serious problems. They could not keep up with the constant influx of war dead.

9. The author, along with his brother Tom, scanned the original hand-written registry for John's name. It could not be found. Other names from the 4th Texas are recorded.

Portion of Oakwood Cemetery taken during the Civil War
Richmond Historical Association

Even as late as 1864, serious allegations were made about the deplorable situation.

From the Richmond *Enquirer*, 6/11/1864

HOLLYWOOD CEMETERY.[10] – This extensive cemetery is fast becoming thickly populated with the dead of our gallant soldiers who expire in the Richmond hospitals from the casualties incident to battle. Many thousands of as brave men as ever handled musket or drew sword in defence of home and country, sleep the last and dreamless sleep under the shade of the grand old oaks with which the cemetery abounds. In later years pilgrims from afar will come to view this spot, for swift, speeding time will throw around it holy memories, to be kept forever green in the hearts of the relatives and friends of those who sleep beneath its soil. We regret that the gentleman in charge of this place is not allowed, during such times as the present, sufficient auxiliaries so that he can satisfactorily discharge the duty with which he is entrusted. Frequently of late, as we understand, have there been such an accumulation of

[10] Hollywood is another major cemetery on the west side of Richmond; Oakwood is on the East side.

A recent photo showing the most likely location of John Tatum's grave which is near the base of the tree on the right, Section A, Row N, near the Confederate Memorial, Oakwood Cemetery

bodies in the evening that no chance was afforded of their having immediately the rights of sepulture. In consequence of this the bodies lie exposed all night, without any protection save that afforded by the rough and generally green pine coffins in which they are brought to the ground. The pine being unseasoned the effect of the hot sun upon them and the fast decomposing bodies within may be imagined. – Frequently the coffin lids, that are only fastened by one nail in each end, burst open, or shrink and curl up so as to expose the body for whose protection it was made.

Several ladies who went to the cemetery, a few mornings since, to mark the grave of a deceased soldier, saw twenty-five coffins, with their ghastly contents, lying on the ground, unburied, and fast decaying under the rays of a burning sun. Myriads of flies were attracted to the spot, and the air was filled with the odors of putrescent flesh. A view of many of the poor fellows was obtained, the lids of the coffins having come off. None of them appeared to have been cleansed and properly prepared for the grave, but thrust in the coffins just as they died. This latter was the fault of the hospital from which

> they came. The fact that the superintendent of the grounds is only allowed four hands (not half enough) is somebody else's fault. The bodies of our dead soldiers should be decently buried. You may use a man and be just, but after using him, you cannot abuse him and be just.

Word was sent back to Texas announcing John's death. It is likely that a local paper would have recorded the deaths of all Texas soldiers from their region. John's son, James, who was three at the time of his father's death, grew up knowing very little about his father. As a result he was not able to pass on to John's grandchildren the details of John's service. The grandchildren never knew how John died, where he died, cause of death, where he is buried, the unit to which he was assigned, or the battles he fought during his few brief months as a Confederate soldier.

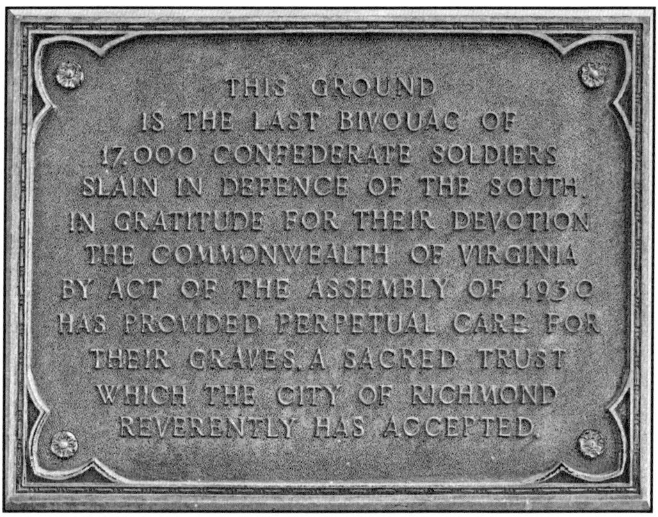

Plaque near the entrance to the Confederate portion of Oakwood Cemetery

Chapter Eight

The War Rages On

The history of the 4th Texas Infantry Regiment after John Tatum's departure for the hospital and eventual death reads a lot like the early months of the war. Chronicled below is a summary of the 4th Texas from November of 1862 to the end of the Civil War in 1865.

November 1862

On November 6, Lee's army was formerly organized into two corps led by Longstreet and Jackson; both were promoted to lieutenant general. John Bell Hood was promoted to major general and his small division was increased by two brigades. The 1st and 4th Texas continued under the commands of Lt. Cols. P. A. Work and B. F. Carter, respectively.

On November 19, the Texas Brigade left its Cedar Mountain camp and marched as the rear of Longstreet's force toward Fredericksburg. They arrived at Fredericksburg on November 22. Despite a 60-mile march over muddy roads, the brigade completed the movement in four days.

Longstreet's command occupied the heights west of Fredericksburg and south of the Rappahannock. Hood's Division was stationed on the Confederate right near Hamilton's Crossing on the Richmond, Fredericksburg, and Potomac Railroad, some four miles south of town. Upon the arrival at Fredericksburg of Jackson's Second Corps from the Shenandoah Valley on November 30, Longstreet concentrated his command closer to

Fredericksburg. Hood was ordered to move his division from Hamilton's Crossing to a position within two miles of the town.

December 1862

The Texas Brigade spent the early part of December drilling, picketing the Rappahannock, and building breastworks along its position near Fredericksburg. The weather was cold, and picket duty -- without adequate clothing and footwear -- was miserable. Burnside's army was poised for attack at any time.

The much anticipated movement by Burnside began on the night of December 10. Pontoons were laid across the Rappahannock, and in two days the Union army was across the river and occupying Fredericksburg. The Battle of Fredericksburg began about 11 AM on December 13 with the attack of Union General William B. Franklin's Grand Division against General Stonewall Jackson's Second Corps on the Confederate right. After an initial success in breaching the position held by A. P. Hill, the Federals were quickly repulsed by Confederate infantry reserves and a murderous artillery barrage. In the afternoon, Burnside launched his main attack against General James Longstreet's left, which was positioned behind a stone wall lining a sunken road at the base of Marye's Heights. Burnside's piecemeal attacks against Longstreet's impregnable position resulted in total failure with heavy casualties. It was pure slaughter. By afternoon, the fields in front of Marye's Heights were covered with frozen dead and dying Union troops. Burnside called off the attacks at sundown. The next morning, Burnside made a feeble renewal of the attack against Longstreet, but it was not pressed. Both sides held their lines until the night of December 15, when Burnside re-crossed the Rappahannock and returned to the safety of his camps at Falmouth. Although Hood's Division occupied the center of the Confederate line, its position was the only one not attacked by Burnside.

The coming of cold and inclement weather soon ended the possibility of continued hostilities. The Texas Brigade staked out a camp among hills and pines just north of the Massaponax River, about a mile in the rear of the line they occupied during the battle. On December 20, the brigade began to build winter quarters of varying types -- wood-framed huts packed with mud and topped with a canvas roofs were typical -- to stave off the inevitable attacks of Old Man Winter.

March 1863

Hood's division changed its base to a camp 100 yards from the Richmond and Petersburg Railroad, the Texas Brigade enjoyed easy access to both the streets of Richmond and the Confederate supply depots. As in the winter of 1861—1862 when John Tatum was recruited, officers and men from the companies of the Texas Brigade were detached back to their home counties in search of men to fill the depleted ranks. As of March 9, 1863, the strength of the Texas Brigade was 3,030 men, only half of which were fit and present for duty. Unlike the previous winter, however, the recruiting drives were not successful. On March 16, General Longstreet received word from General Lee that Major General Joe Hooker, who was now commanding the Union Army of the Potomac, appeared ready to move south across the Rappahannock. On March 18, the Texas Brigade marched northward along the Richmond and Petersburg railroad.

April 1863

When it became clear that a Federal advance on Richmond from the southeast was not planned, Lee changed the mission of General Longstreet's two divisions. Hood's and Pickett's men were now to join Confederate forces in southern Virginia and pin down the Federal garrison at Suffolk, Virginia, near the Great Dismal Swamp. Suffolk was well fortified and occupied by 25,000 to 30,000 troops under Federal commander, Major General John J. Peck. A flotilla of Union gunboats patrolled the Nansemond. Longstreet's 20,000 men entrenched along a 15-mile line around the town, from the Nansemond in the north to the Great Dismal Swamp in the south. Hood's Division occupied the left or northern wing. The commander of Co. G, 4th Texas, reported that his unit was "engaged in action with 3 gunboats on the Nansemond River on the 14th and 15th of April."

May 1863

General Longstreet issued orders to his commanders to leave the Suffolk area and proceed without delay to rejoin the main body of the army near Fredericksburg. Hood's Division -- the last to leave the Suffolk trenches -- assumed its usual position as the rear guard of Longstreet's command. By the time Longstreet's command had left the Blackwater region, General Lee had fought and won the Battle of Chancellorsville. As Longstreet's presence was no longer urgently needed, Lee ordered Longstreet on May 7 to proceed north at a normal marching pace. The Texas Brigade reached Richmond on May 8, and was given a handkerchief-waving welcome by the ladies of the capital city.

June 1863

On June 3, General Lee commenced his second great invasion of the North by withdrawing his army from the Rapidan-Rappahannock line and moving it toward the Shenandoah Valley.

The Texas brigade began a long, slow march toward the Shenandoah Valley, then north toward Maryland and Pennsylvania. On June 24, General Robertson marched the men into Maryland. The Texans were in high spirits, and were looking forward to sampling the bounty of Northern territory. June 26, 1863, Hood's men performed a feat never again matched by any division in the war. They had breakfast in Virginia, lunch in Maryland, supper in Pennsylvania, and slept in a state of intoxication -- four states in 24 hours. The next morning, June 27, the Texans resumed their march north, passing through Greencastle up the Cumberland Valley toward Chambersburg, Pennsylvania. As the men passed through the streets of Chambersburg, many of the townsfolk wore patriotic banners and made derisive remarks about the ragged Confederates. One woman wore a large American flag draped across her ample bosom until a Texan hollered out, “Take care, Madam, for Hood’s boys are great at storming breastworks when the Yankee colors is on them!” Lt. Col. Freemantle noted that the women beat a “precipitate retreat.”

July 1863

Early in the morning of July 1, the Texas Brigade left its bivouac at Fayettesville and resumed its march along the Chambersburg Pike toward Cashtown, which lay 12 miles to the east. The Texas Brigade departed from Cashtown about 2 AM on July 2. There the men were permitted to stack arms and rest. Word of the Confederate victory on July 1 in the fields north and west of Gettysburg must have reached them by this time. The men had rested but two hours in Cashtown when they received orders to resume their march toward Gettysburg, which was 8 miles further east along the pike. The Texas Brigade reached General Lee’s headquarters, just west of the town and south of the pike, an hour after sunrise of July 2. After a short delay, the Texans moved about a mile southwest to the valley of Willoughby Run behind Seminary Ridge.

The Federals had been beaten back in the previous day’s fight, but had assumed a strong defensive position on Cemetery Ridge, Cemetery Hill and Culp’s Hill, south and east of Gettysburg. Lee wanted an echelon attack against the Federal left flank by Hood’s and McLaws’ Divisions of Longstreet’s Corps and by Richard Anderson’s Division of A. P. Hill’s Corps. Longstreet opposed Lee’s plan and advocated maneuvering around the Federal right flank, placing the Army of Northern Virginia between the Army of the Potomac and Baltimore, and receiving a Federal attack on ground of their own choosing. Lee viewed Longstreet’s plan as impractical and risky, and ordered the attack against the Federal left to begin as soon as practicable.

By 4 PM, McLaws and Hood finally had their divisions in place for the attack. Hood’s Division deployed in two lines of two brigades astride the Emmitsburg Road about 2 ¾ miles south of the Lutheran Seminary. The Texas Brigade was on the left of the forward line and Law's Brigade was on its right. From left to right were deployed the 3rd Arkansas, 1st Texas, 4th Texas, and 5th Texas. Despite efforts to conceal their arrival, a hail of hot iron from Federal batteries in the Peach Orchard and Devil's Den made it plain that the Federal's knew their position.

Soon after reaching his assigned position, Hood received word from Law's scouts that the Federal left flank extended no further south than the Round Tops. Hood was convinced that turning the Round Tops would permit an easy attack of the enemy's rear. He petitioned Longstreet via courier to allow him to make the maneuver, but Longstreet denied the request by quoting General Lee's orders to advance up the Emmitsburg Road. Hood petitioned twice more -- the last time in person -- further stating that the rocky terrain to his front precluded an unbroken advance on the enemy position. Hood's petitions were each denied by Longstreet and Hood formally protested the attack.

Having received his orders from Longstreet to execute Lee's plan, Hood ordered his brigades to attack. It was just after 4 PM. Hood, on horseback in front of the Texas Brigade, commanded, "Forward-Steady-Forward!" Hood ordered Robertson to keep his brigade in contact with the Emmitsburg Road on the left *and* Law's Brigade on its right. Law's Alabamians, who received a head start in the attack, soon outpaced the Texas Brigade, and Robertson's men had to break into a trot just to keep contact on their right. Robertson quickly realized that he would have to violate Hood's order by either losing contact with Law or the road. Robertson and Col. Van Manning of the 3rd Arkansas opted to break from the road rather than split the division.

Soon after crossing the Emmitsburg Road, Hood left Robertson and rode to the center of the division. While there, a shell exploded above his head, and one of its fragments tore into his left arm. Hood reeled in the saddle from shock, and his staff and comrades lowered him to the ground. Stretcher bearers carried him from the fight that had barely begun. Col. Evander Law then assumed command of Hood's Division.[1]

The 4th and 5th Texas advanced by the Bushman and Slyder farmyards, splashed through Plum Run, and entered the woods on the western slope of Big Round Top. As the 4th Texas advanced toward a stone wall, riflemen from the Second US Sharpshooters posted behind it poured a destructive fire into the regiment's flank. Here, Lt. Joe C. Smith was killed, and Col. John C. G. Key and Lt. Col. Benjamin F. Carter (the former captain of Co. B) were wounded. Carter died a few days later in Federal hands. To the 4th's right, the 5th Texas and 4th Alabama cleared the sharpshooters from the wall and drove them back. The three regiments then pressed on through thick woods and boulders to the saddle between the Big and Little Round Tops.

When the 4th Texas passed to the east of Devil's Den, the men on its left could see the fight for Smith's Battery being waged. Cpl. Miles V. Smith of Co. D assembled several of his fellow 4th Texans and sniped at the cannoneers from the boulders on Big Round Top. Smith was rewarded by the sight of the 1st Texas and the 20th Georgia of Benning's Brigade capturing the battery. When Federal troops attempted to recapture the guns, Smith's detail fired into the Federal flank. When the fight ended, he found the 4th Texas halfway up the slope of Little Round Top.

Due to the large presence of Federals on the top of the crest, the two Texas regiments retreated back down the slope of Little Round Top. Ordered forward, they advanced a second time. By this point, the Texas and Alabama regiments were attacking in only loose coordination at best. With most of their company officers dead or wounded, the men attacked through the boulders as squads instead of regiments. Men were everywhere trying to take cover behind the boulders. Inevitably, the second assault end-

1. The division surgeon decided not to amputate Hood's left arm although that would have been the standard approach. The doctors had no way of pinning the bones back into place. Hood's arm hung useless at his side for the rest of his life.

ed just as the first. Joined by the Forty-eighth Alabama on their left, the 4th and 5th Texas attempted a third assault. As darkness descended, the noise of the battle faded into quiet. Night, however, brought little relief to the Texans wounded or stranded behind the boulders and in the crevices of Little Round Top. Scouting parties from both sides retrieved the wounded or captured prisoners. Federals and Confederates alike hastily built stone walls in anticipation of renewed battle in the morning.

At daylight on July 3, the scattered remnants of Robertson's Brigade were unified and occupied a sector along Plum Run between Devil's Den and Big Round Top. This would become part of the main Confederate defense line throughout the day. While Lee was launching "Pickett's Charge" against the Union center, the Confederate right was relatively quiet. Late in the afternoon, Hood's Division withdrew from its forward position to a line near the Emmitsburg Road. Here they remained through July 4, awaiting a Federal attack that never came. The Battle of Gettysburg was over.

The Texas Brigade suffered the second highest brigade loss in Hood's Division -- 84 killed, 393 wounded, and 120 missing. Heavy casualties were suffered in the top officer ranks: General Robertson, Cols. Manning, Key, and Powell, and Lt. Cols. King Bryan and Benjamin Carter were all wounded, the last fatally.

September 1863

On September 2, General Evander Law, commanding the division for the injured General Hood, ordered General Robertson to move the Texas Brigade about 20 miles below Fredericksburg to Rappahannock Academy near Port Royal. Robertson was instructed to post pickets at all river crossings and report any attempted crossing by the Federals, a sizeable number of whom had recently been spotted in the area.

The Texas Brigade remained near Port Royal until September 7, when it received orders to break camp and march to Milford Station. Once there, the brigade (minus

Reilly's North Carolina Battery) was to board the cars of the Richmond, Fredericksburg, and Potomac Railroad bound for the Confederate capital. This order was part of a grand movement to send Hood's and McLaws' Divisions west to Tennessee to join General Braxton Bragg's Army of Tennessee, which opposed a Federal army under General William S. Rosecrans near Chattanooga. Hood, pressed by his men, joined them for this move south.

The Battle of Chickamauga began at dawn far on the Confederate right. By mid-afternoon, the Confederate left was engaged. Between 3 and 3:30 PM, the Texas Brigade was ordered forward to assist the hard-pressed Johnson. As the Texans moved forward, they encountered many stragglers from Johnson's Division streaming to the rear. With Sheffield's (Law's) Brigade to their right, the Texas Brigade emerged from the heavy underbrush and woods and marched into a clearing.

During this advance, the Texans saw General John Gregg of Johnson's Division, who had been shot through the neck and dragged by his horse's reins toward the Texas Brigade, lying in no-man's land between the enemy lines. Gregg's spurs and sword were being pilfered by some adventurous Federals when the Texas Brigade charged forward, drove off the Yankee scavengers, and rescued Gregg and his horse. Little did the Texans know that within a few months Gregg would be commanding their own brigade.

As the Texas Brigade advanced some 200 to 300 yards into the clearing, it was viciously attacked from the left by Federals posted in a ravine covered by thick underbrush. Changing their front and leaving Sheffield's supporting brigade, the Texas Brigade attacked and drove the Federals from the ravine and onto the high ground beyond. Here Robertson's men came under frontal fire from barricaded infantry on the crest and assailed from the right by two Federal batteries firing grape and canister at 200 yards. Val Giles of Co. B, 4th Texas recalled that a group of Federal prisoners being herded to the rear was hit by the artillery, and stated that he "distinctly saw the dust rise from [one Federal prisoner's] shoulders where the grapeshot struck him." The Texas Brigade rushed forward across the Lafayette-Chattanooga Road and closed with the enemy. Though some of the Federals fled their barricaded position, many remained behind the rail fences and engaged the Texans and Arkansans in fierce hand-to-hand combat before abandoning the hill.

The bitter fighting shifted toward the Viniard House and Farm, where Federals had tenaciously beaten back Johnson's previous assaults. The Texas Brigade recklessly charged the fortress-like log house, eventually forcing the enemy from the "hornet's nest" but with great loss. The 4th Texas suffered the most casualties in the attack, including Lt. Col. Bane, Lts. J. M. Bookman and A. G. Killingsworth, and Color Sergeant Ed Francis. The regiment would eventually bury 22 of its men on the grounds of the Viniard Farm.

The driven Federals deserted several pieces of artillery and fell back to a skirt of timber a quarter of a mile west of the Lafayette-Chattanooga Road. Unable to cope with the superior Federal numbers and firepower, the Texas Brigade was forced to withdraw from its salient back toward the main Confederate line. As they fell back through the farm and beyond the hill from which they had previously driven the Federals, the Texans again came under heavy and accurate artillery fire. Federal infantry pursued the Texas Brigade during its retreat and reoccupied the hill. In a desperate charge, the Texas Brigade drove the Yankees from their barricaded position for the second time. Robertson, anxious to hold onto the only high ground in the area, ordered artillery and infantry support. To this request, he received only Benning's infantry. The two brigades managed to repel repeated assaults and hold the hill until sunset. After three requests for artillery support went unanswered, Robertson and Benning were ordered by General Law to withdraw their commands under cover of darkness 150 yards back across the Lafayette-Chattanooga Road.

About 11AM on September 20, Hood ordered all his divisions forward simultaneously toward the Lafayette-Chattanooga Road. The Texas Brigade crossed the road just south of the Brotherton House and advanced quite a distance through broken, wooded country. After advancing about a mile, much of it under artillery fire, Robertson's men were fired upon by a strong Union force of infantry and artillery posted on a wooded hill to their right front. The Texas Brigade immediately wheeled to the right and moved against the hill. The men pushed forward through the underbrush at the quick step, firing as they went. With accurate and heavy fire, the Texas Brigade drove the Federals from the hill, and caused them to retreat in disorder to a second ridge a short distance away. From this position, the Federals maintained a steady fire against Robertson's command.

The Texas Brigade had advanced so rapidly to its front and right that it was well ahead of the Confederate army in that section of the field. This rapid advancement was due to a Federal blunder in which Rosecrans had ordered General T. J. Wood's Division to move from the right to the center of the Federal line about the same time that Longstreet and Hood launched their attack. Eight Confederate brigades, including the Texas Brigade, poured through the resulting half-mile gap in the Federal line.

As the Texas Brigade withdrew from its position in some confusion, General Hood happened to be riding by as he was directing the advance of his very extended command. He saw his old brigade retiring, and, seeking to rally them, galloped over to the timber where they were reforming. As Hood and Robertson were conversing in front of the woods, a minie ball struck Hood in the upper part of his right thigh, splintering the bone. Hood fell from his horse into the arms of his old brigade. Several Texans who witnessed the shooting thought that the shot came from the same friendly troops, perhaps from Florida, that had been firing upon them before. Hood was carried from the field after ordering Bushrod Johnson to press the attack. His leg was amputated shortly after.[2] This day would be the last the Texas Brigade would ever serve under the command of Major General John Bell Hood. In the two days of fighting at Chickamauga, the Texas Brigade lost 570 of about 1300 men.

November 1863

After many days of skirmishing near Chattanooga, the Texas Brigade, on November 23 and after much long-range skirmishing and under artillery fire, opened the attack on the main Federal line near Knoxville. The Yankees resisted stubbornly at first, but they eventually gave way in disorder and retreated to a prepared defensive position on the crest of a high ridge. For the next two days, the Texas Brigade kept up a steady fire against the strong Federal position, but failed to drive the enemy from it. General Robertson then withdrew his command to Cherokee Heights, a series of high hills about 1000 yards west of Fort Higley, and set up a defense line. From this position, the Texas Brigade exchanged sniper fire with the Federals at Fort Higley and across the river.

December 1863

On December 2, General Longstreet issued orders to withdraw from the vicinity of Knoxville. The order was in response to a telegram from President Davis received four days earlier informing Longstreet of the Federal victories against General Bragg at Lookout Mountain and Missionary Ridge. Bragg had retreated into northwestern Georgia and Longstreet was to make an attempt to rejoin him. Logistics, however, precluded a march across the mountains to Georgia, so Longstreet instead ordered a retreat to Bristol, Virginia, where he planned to encamp for the winter. The route of retreat was north around Knoxville and then northeast along the north bank of the Holston River.

---

2. After recovering from his latest wound, Hood continued to serve throughout the remainder of the war. With only one working arm, a leg missing, his men had to tie him to his saddle so that he could ride into battle.

On December 3, Law's and Robertson's Brigades vigorously attacked the Federals to their front in an attempt to conceal the withdrawal. After withstanding the Confederate attacks until noon, the Federals retreated to a second line of entrenchments nearer the Holston. The two Confederate brigades then quietly withdrew westward, crossed to the north bank of the river by ferry in the early evening, and marched north around Knoxville.

On December 10, Longstreet received another telegram from Davis informing him that he had been given sole authority over the troops in his Department of East Tennessee. That same day, General Robertson wrote to General Hood that his brigade carried on its rolls only 784 men "present for duty," of which "many" were "not fit to march." Thus, the Texas Brigade now had the effective strength of an undersized regiment. Of the "present for duty", the 4th Texas counted 20 officers and 193 men -- the most of the four regiments in the brigade. Robertson proposed to Hood that the Texas Brigade be sent back to Texas for the winter to recuperate and recruit and then rejoin Longstreet west of the Mississippi on or about April 1, 1864. Robertson's proposal was ultimately refused by the War Department.

On December 18, General Micah Jenkins preferred charges of "conduct highly prejudicial to good order and military discipline" against Robertson. These charges stemmed from detrimental remarks that Robertson had supposedly made about Jenkins' generalship and judgment following a recent order to pursue marauding Federal cavalry. According to Miles V. Smith of the 4th Texas, the "Brigade was called on often during the winter to race out after Federal cavalry over ice and snow with no shoes." When ordered to repel a cavalry raid under bad weather conditions, Robertson protested and remarked that "he didn't see any reason to have his men run after every cavalryman that came by." A "junior officer" reported this remark to Jenkins.

On December 23, Longstreet relieved Robertson of command of the Texas Brigade and ordered him to Bristol,

Tennessee, to "await the assembling of a court for the trial of his case." Lt. Col. King Bryan of the 5th Texas assumed temporary command of the brigade. (Bryan would later be replaced by Captain A. C. Jones of the 3rd Arkansas.)

January 1864

The winter of 1863-1864 was harder on the Texas Brigade than the previous two. More men than ever were without shoes in the freezing weather. A trail of blood followed the men from Texas when they marched over the frozen roads. The hardships experienced by the Texas Brigade at this time were exacerbated by their concern over the fate of General Robertson, who was currently under arrest at Bristol and awaiting trial by court martial. On January 6, all officers present for duty in the 4th and 5th Texas signed a petition to the Secretary of War asking that General Robertson be restored to the command of the brigade. About the same time, the 3rd Arkansas forwarded to Richmond a similar petition. No such petition was sent by the 1st Texas, which was without a field grade officer at the time.

February 1864

On February 3, a court martial was at last convened at Russellville, Tennessee, to review the charges against General Robertson that were preferred by General Jenkins on December 18, 1863. On February 12, President Davis finally settled the dispute between Jenkins and Brigadier General Evander Law over the command of Hood's old division by assigning the command to Major General Charles W. Field. (Field was a jovial Virginian who was seriously wounded at the Second Battle of Manassas and had recently returned to the army as an amputee.) Both Law, who had resigned over the dispute in December, and Jenkins were reassigned to the commands of their old brigades in Hood's, now Field's, Division.

The miseries of wintering in East Tennessee continued. By this time, Longstreet's hungry and scantily clad men were further victimized by a severe infestation of lice. Not since their first winter at Dumfries, Virginia had the lice been so bad. From the various personal accounts describing the parasites, they were a larger species than those the men remembered back in Virginia.

Meanwhile, the court martial of General Robertson had reached a decision in Russellville. On February 25, the court acquitted Robertson of "improper motives" but disapproved "his conduct." The court reprimanded Robertson and permanently relieved him of command of the Texas Brigade. He had commanded the Texas Brigade longer than any other officer. Robertson was replaced by his judge and fellow Texan, Brigadier General John Gregg. (Gregg's old brigade was eliminated in the reorganization of the Army of Tennessee in late September 1863.) After his trial, Robertson went to Richmond to await a new assignment.

March 1864

For most of March 1864, the Texas Brigade remained in winter camp at Bull's Gap enduring the lingering harsh winter, chronic shortages of food and clothing, and an infestation of lice. After inspecting the Texans in early March, Major C. M. Winkler of the 4th Texas noted that the greatest need of the men was underwear. This dire need was made known to the public by means of letters sent to various relief agencies and newspapers throughout the South. Moved by the plight of the Texas boys so far from home, the ladies of Virginia and Georgia in particular shipped in boxes of underclothing, which Winkler reported as being "very acceptable."

April 1864

During late March and early April the Texans were sent in the direction of Virginia. They were elated that they were finally leaving the miseries of Tennessee. On April 9, just before leaving Richmond for Texas, General Robertson wrote a poignant farewell to his old brigade. The last sentence was, "With a mind saddened by the remembrances of ties broken, and with the prayer that God, in his mercy, will guard, protect and bless you, I bid you farewell." With this, Robertson left the Army of Northern Virginia and his beloved troops.

On April 11, General Longstreet received orders from Richmond to report with the "original portion of the First Corps to General R. E. Lee." The Texans and Arkansans boarded the train at Zollicoffer on April 20. They arrived in Lynchburg, Virginia, on April 22 and then continued on to Charlottesville the same day. On April 26, the Texas Brigade marched north 16 miles to Cobham Station (Depot) on the Virginia Central Railroad. Here, about 8 miles below Gordonsville, the brigade went into camp and remained for the remainder of the month.

May 1864

On May 1, the Army of Northern Virginia was deployed on a wide front, with the First Corps (Longstreet's) near Gordonsville on the Orange and Alexandria Railroad, the Second Corps (Ewell's) along the Rapidan River near Mine Run, and the Third Corps (A. P. Hill's) posted on Ewell's left further up river. On May 4, the 120,000 man Army of the Potomac, commanded by Major General George Meade, but closely supervised by Commander-in-Chief Lt. General Ulysses S. Grant, crossed the Rapidan at Germanna and Ely's fords and commenced its long awaited spring offensive against Lee and Richmond. Lee, who had anticipated the crossing since May 2, did not dispute Grant's and Meade's movement. Instead, he let the Army of the Potomac march into the thick, tangled forest known as "the Wilderness," where the Federals' advantages in artillery and manpower would be greatly diminished.

Longstreet was ordered to the area of the Wilderness to counter the Federal movements. When Longstreet's column reached the Orange Plank Road and turned east they heard the sounds of the furious battle coming from the east. "Thinking the emergency might be great," Field drew the head of his division alongside and 100 yards to rear of Kershaw's lead brigade. From then on, the two divisions marched in parallel columns of four along each side of the road, with Field's on the left and Kershaw's on the right. Instinctively, the leading brigades -- the Gregg's Texas Brigade of Field's Division and Humphreys' Brigade of Kershaw's Division -- increased their pace until a race toward the battle had begun between them.

Longstreet immediately gave the order for his men to advance the last mile and a half at the double quick. Reaching the Widow Tapp's farm, Kershaw's Division began forming for battle south of the Orange Plank Road. Before Field could follow suit, Longstreet ordered Field to deploy his front north of the road. Field acted with dispatch and ordered Gregg's 800-man Texas Brigade to execute a right wheel into line perpendicular to the road and behind Poague's guns. The Texas Brigade was arranged in its usual battle order -- from left to right, 3rd Arkansas, 1st Texas, 4th Texas, and 5th Texas. These regiments were commanded by Col. Van Manning, Lt. Col. Frederick Bass, Col. John P. Bane, and Lt. Col. King Bryan, respectively. Behind the Texas Brigade were stacked, in order, Benning's, Perry's (Law's), and Jenkins' Brigades.

As Gregg ordered the Texas Brigade forward through Poague's guns, a much relieved and excited General Lee rode toward him and asked, "General, what brigade is this?" "The Texas Brigade," Gregg proudly answered. Lee replied, "I'm glad to see it! When you go in there, I wish you to give those men the cold steel. They will stand and fight all day, and never move unless you charge them." When Gregg received Longstreet's order to advance, Gregg saluted Lee, rode to the center of his command, and boomed for all to hear, "Attention, Texas Brigade! The eyes of General Lee are upon you! Forward...march." Lee could not contain his excitement. Rising high in his stirrups, and waving his hat, he shouted, "Texans always move them!"

After Lee's exhortation, "a yell rent the air that must have been heard for miles around, and but few eyes in that old brigade of veterans and heroes of many a bloody field was undimmed by honest, heartfelt tears." Lee, moved greatly by this response, moved through an opening in the brigade and attempted to lead the Texans in their charge. When the Texans realized this meant almost certain death for Lee, a cry of "Go back, General Lee. Go back!" spread across the entire line. As Lee urged his horse Traveler on, several men nearest him sprang from the ranks and grabbed at the reins and saddle. Hearing calls of "General Lee to the rear" and "We won't move until you go back", Lee finally acquiesced, but not before Major Venable and General Gregg "forcibly" persuaded their leader to leave the field and join Longstreet on a knoll near Poague's artillery.

Texans and Arkansans charged, at the double quick, into the first line of Federals. That blue line broke, and the Texas Brigade streamed forward -- unsupported -- into the main Federal line in the woods beyond the Widow Tapp's farm. There came a terrible crash of musketry, as Gregg's severely outnumbered troops exchanged a brutal, stand-up fire with the enemy no more than 20 yards away. "Death seemed to be our portion," one Texan later wrote.

The Texas Brigade charged again, and the main Federal line bowed back to a second line of breastworks. Face-to-face fighting ensued. During the charge, a strong enemy force posted south of the Orange Plank Road opened heavy enfilade fire on the 4th and 5th Texas. Gregg ordered the two regiments to swing right across the road to attack their assailants. Union artillery posted in the road raked the regiments with double-shotted canister as the line of Texans moved diagonally across their front. The 4th and 5th Texas moved steadily forward until coming within 100 yards of entrenched Federal infantry south of the Orange Plank Road. Learning that a large Federal force was advancing down the road and fearing the regiments would be cut off, Gregg led the 4th and the 5th back north of the road.

"For 25 minutes we held them steady," boasted a Texan who lived through the carnage, "and at the expiration of that time more than half our brave fellows lay around us dead, dying, and wounded, and the few survivors could stand it no longer." Gregg was nearly killed as his horse was shot beneath him. At his command, the brigade grudgingly gave ground. Only 250 men were able to follow Gregg's order and return to the Widow Tapp's farm. These remaining men reformed about 300 yards in front of Poague's artillery and replenished their ammunition.

Although no official casualty reports for the Texas Brigade survive, the Galveston *News* reported on May 7, 1874, that Gregg's command lost 565 of 811 men that went into action on May 6, 1864. If correct, this 70% casualty rate would be the highest incurred by the Texas Brigade during the war. According to Miles V. Smith of Company D, the 4th Texas had 207 men present for duty at the battle and lost 130, including four color bearers. All the officers of the 5th Texas were lost. Col. Manning of the 3rd Arkansas was badly wounded and later captured. Afterwards, Gregg confided to Postmaster General John Reagan that if Longstreet had not recalled his men "he believed the old brigade would have been annihilated, for in his opinion the Texans had not intended to go back alive."

Perhaps the most significant of Lee's losses during the Battle of the Wilderness was that of Longstreet, who, during a flanking maneuver later in the day, was shot along with several of his staff and brigadiers by unwitting Confederate troops.

Longstreet was seriously wounded in the neck and Brigadier General Micah Jenkins was killed outright. Longstreet would later return to his command after a period of recuperation. The incident of friendly fire was eerily reminiscent of the fatal wounding of Stonewall Jackson almost exactly a year before on the other side of the Wilderness at Chancellorsville.

At 2 PM, augmented by Brigadier General John Gibbon's Division from Hancock's Corps, Warren commenced his second attack of the day. Gibbon's 1st Brigade, under Brigadier General Alexander Webb, assaulted the formidable abatis and breastworks constructed by the Texas Brigade. Supported by the Richmond Howitzers, the Texans and Arkansans tore bloody gaps into Webb's line, and forced the Federals to lie on the ground. Pine needles around them ignited, and Webb's men beat out flames with their caps as they crawled on their bellies to escape Gregg's bullets. The story was the same all along the Federal line, and after an hour, Warren called an end to this latest disastrous attack.

At 5 PM on May 10, Grant ordered a full-scale attack against the now well-fortified Confederates. The attack against Field's sector was begun at 7 PM by Hancock's Corps. The Texas Brigade was busy preparing its supper when it caught sight of the brigade of Brigadier General Hobart Ward marching at the double quick toward their works. Jumping for their guns and smashing open fresh boxes of ammunition, Gregg's men waited for the Yankees to reach their position. Once reached, the Texans and Manly's North Carolina Battery soon began raking the Yankees with bullets and canister. The 4th and 5th Texas and the 3rd Arkansas managed to hold the Federals in check, but the 1st Texas could not prevent the bayonet-wielding Yankees from breaching their works via a 40-foot wide gap over a gully. Bloody hand-to-hand fighting ensued. The Texans, lacking the bayonets they discarded in Tennessee, wielded their rifles as clubs. As the Federals poured through the breach, they were met with enfilade fire from the 3rd Arkansas on the left and the 4th and 5th Texas and "Tige" Anderson's Brigade on the right. Unable to stand the flanking fire, the Yankees withdrew in confusion through the breastworks. As they fled in disorder back to their lines, the Federals suffered severely from artillery and small arms fire.

Anticipating another sidling maneuver by Grant around his right flank, Lee prepared his troops for a move southward. Stationed on Lee's left, Field's Division would be the first to move. On the night of May 14, the Texas Brigade received its movement orders. At noon on May 15, Gregg's men marched southeast across the rear of Lee's army to a position south of A. P. Hill's Corps along the Brock Road. Here they remained until May 20, when Grant began withdrawing his 5th and Sixth Corps from the Spotsylvania lines. The Texas Brigade was ordered to go south of the Po River "in support of the guns there," and to connect with Ewell's Corps moving down from Spotsylvania Court House.

On the afternoon of May 21, the Texas Brigade made a movement south on the Telegraph Road with Field's Division. The brigade crossed to the south side of the North Anna River and, after marching a mile, halted near Hanover Junction on the Richmond, Fredericksburg and Potomac Railroad. By the night of May 22, Lee's entire army was below the North Anna. As expected, Grant appeared on the north side of the river the next morning. Both armies hastily dug in and prepared for battle. Between May 23 and 26, Grant launched a series of simultaneous probing attacks against the flanks of Lee's strongly entrenched army.

June 1864

Grant continued to move southeast toward the Chickahominy River. Lee followed on a parallel route. By May 31, both armies were approaching Cold Harbor, near the old battlefield of Gaines' Mill -- where Hood's Texas Brigade had first gained fame two years before.

By June 1, the Union and Confederate armies confronted each other at Cold Harbor in positions opposite to those they assumed two years before. Instead of facing southeast as they had during the Battle of Gaines' Mill, the Confederates now faced northeast. Lee's line extended from Totopotomoy Creek on the north to the Chickahominy River on the south -- a distance of eight miles. Ewell's Corps occupied the left wing of the army, Anderson's Corps the center, and A. P. Hill's corps (commanded by Jubal Early) the right. By remarkable coincidence, the Texas Brigade took up a position on the same ridge from which it had helped drive Fitz John Porter's Corps on June 27, 1862.

Gregg's men were up early on June 3, many of them breakfasting on "clammy corn-bread and raw bacon", when Grant struck in force at 4:30 AM. Some 40,000 Federals from the corps of W. S. Hancock, Horatio Wright, and "Baldy" Smith advanced first against A. P. Hill and then Anderson. The men of Smith's Eighteenth Corps assaulted the position held by General Evander Law's Brigade of Alabamians. The Texas Brigade, posted to Law's right, poured a destructive enfilading fire into the ranks of Smith's advancing men.

All along Lee's line, the Federal assault withered under deadly Confederate fire. Double-shotted canister swept away entire Federal companies with each barrage. In a little over eight minutes, Grant lost over 7,000 men. The grand assault was a dismal failure and won for Grant the sobriquet, "The Butcher." After the war, General Law would remark, "It was not war, it was murder." Law counted over 1,000 dead and wounded in front of his works at a loss of not more than 20 in his command.

For the next six days, the opposing armies sniped at each other without a general battle. On June 12, Grant decided that it was impossible to take Richmond by going over or around Lee's army. Changing his objective, Grant withdrew the Army of the Potomac from its position and set it in motion southward across the Chickahominy and James Rivers toward Petersburg. That city, a rail hub located on the Appomattox River 23 miles south of Richmond, controlled a vital supply line to the capital and to Lee.

On the morning of June 18, the Texas Brigade marched west to the Richmond and Petersburg Railroad, where they boarded the cars bound for Petersburg. Late in the day, Gregg's command moved into the trenches east of the city and south of the Appomattox River. The Texas Brigade, on the left of Field's Division, was positioned to the right of Kershaw's Division. The brigade was located mid-way between the Norfolk and Petersburg Railroad and the Jerusalem Plank Road about 1.5 miles east of Blandford Cemetery. The right of the brigade rested on Taylor's Branch. Here Gregg's men would begin a weary month of duty in the trenches.

July 1864

A religious revival that swept through Lee's army in the summer of 1864 had mixed effect on the Texans. O. T. Hanks of the 1st Texas wrote of "a good sermon" given by the brigade chaplain, the Rev. Sam Davis, but doubted the value of the revival, stating that "we are all pretty good Christians by this time and will be until the quiet future...[After the war,] some of us will return to `wallow in the mire." Major Winkler of the 4th Texas reported that a popular pastime in his regiment was singing religious songs. Winkler also noted with pleasure that chess was "fast superseding" cards as a recreational pursuit and that gambling was becoming most "unfashionable."

While in the trenches at Petersburg, the Texas Brigade heard a rumor that the Federals were tunneling and constructing a mine under the Confederate lines. This rumor prompted some ingenious methods of detecting the subterranean diggers. Behind the Confederate entrenchments, pegs were driven deep in the ground. By biting and holding these pegs in his teeth, a Confederate could detect the slightest vibration in the ground. O. T. Hanks reported that he detected no tunneling or mining in the sector of the defenses manned by the Texans.

The Texas Brigade remained in or just behind the front lines east of Petersburg without relief until July 20 - longer than any other brigade. Soon after dark on July 28, the Texas Brigade and the rest of Field's Division left the Petersburg trenches and began an all-night movement to join the rest of Anderson's 1st Corps, two divisions of A. P. Hill's Third Corps, and Fitzhugh Lee's Cavalry Division near Richmond.

At 4:30 AM on July 30, the rumors of a mine were validated as the Federals exploded 8,000 pounds of black powder under a part of the Confederate's Petersburg line known as "Elliot's Salient." The explosion, which created a crater 170 feet long, 60-80 feet wide, and 30 feet deep, was followed by a uncoordinated and unsuccessful Federal assault on the breached Confederate works. Although the Confederates successfully sealed the breach, the Battle of the Crater had been costly. In the explosion alone, some 300 Confederates from Bushrod Johnson's Division and Pegram's Battery were killed, wounded, or otherwise lost. The explosion occurred in the immediate vicinity of the position held by the Texas Brigade at Petersburg until July 28. Had they not been ordered northward across the James, Gregg's men would surely have shared the fate of Johnson's and Pegram's men. In fact, some of the Federal prisoners were reported to have been disappointed because the Texas Brigade was not "extinguished" in the "grand upheaval and collapse."

September 1864

Nearly a month after the unsuccessful attacks against the Richmond-Petersburg defenses at the Weldon Railroad and Fussell's Mill, General Grant again attempted a two-pronged attack against the ends of the Confederate defenses. The primary purpose of this attack was to prevent General Lee from reinforcing Major General Jubal Early's army operating in the Shenandoah Valley. By late September, the Confederate defenses around Richmond were thinly manned by two brigades (Benning's Georgians and Gregg's Texas Brigade) of Field's Division, one brigade (Fulton's) of Edward Johnson's Division, one Virginia militia battalion, and Gary's South Carolina Cavalry Brigade. Protected by an early morning fog and heavy woods, two Federal corps struck the Richmond defenses near Forts Harrison and Gilmer at daylight on September 29. As the assault opened, the Texas Brigade was called upon to defend the outer works near Fort Harrison. The brigade was aligned from left to right by the 1st Texas, 4th Texas, 5th Texas, and 3rd Arkansas. The Texans and Arkansans were attacked by a brigade of Negro troops from Birney's Tenth Corps that charged through an apple orchard near and up a narrow valley near the Phillips House.

The encounter between the Texas Brigade and the Negro brigade was a massacre. The black troops found themselves hemmed in by the banks of a dry creek bed and thus easy targets for the sharp shooting Texans. According to

Joseph B. Polley of the 4th Texas, "Not a dozen shots in all were fired" by the Federals, and "not a man in the Texas Brigade received a wound." In just five minutes, 194 Negro soldiers and 23 of their white officers were killed and several times that number wounded. In places, the dead were stacked five deep. Two Federal regimental flags were captured in the debacle, one by the 4th Texas and the other by the 3rd Arkansas. The Texas Brigade took 43 of the Negro soldiers captive. According to Polley, most of the prisoners preferred to serve their captors as orderlies rather than risk confinement in a Southern prison.

October 1864

Disturbed by the loss of the outer Richmond defenses near Fort Harrison, General Lee immediately began plans to drive the Union forces from their newly won, but strongly fortified, position. Lee planned a frontal attack by the divisions in the immediate area.

The main Federal defenses featured formidable log breastworks atop a hill, a thick abatis 500 yards in front of the breastworks, and a large swamp in front of the abatis. Gregg was ordered to spearhead the assault straight down the Darbytown Road -- precisely where the Federal entrenchments were strongest. Not since Gaines' Mill had the Texas Brigade been asked to storm such a strongly prepared position. The Texans and Arkansans were to be supported by the rest of Field's Division on the left and Hoke's command on the right.

As the Texas Brigade approached within a quarter mile of the Federal position, it came under heavy artillery fire which forced them to take cover and regroup. After a short pause, the brigade advanced forward and waded through the swamp, but had their mass formation fragmented by the thick abatis beyond. Neither Anderson on the left nor Hoke on the right advanced in support of the Texas Brigade, so Gregg's flanks were left exposed to the enemy's enfilading fire. Kautz's cavalry, armed with new Spencer repeating carbines, poured a destructive fire down upon the Texans. A handful of Gregg's men advanced within 30 paces of the blazing barricade where they were pinned down and either captured or shot. A large group found protection in a depression 300 yards short of the breastworks, but others remained tangled in the abatis and became prime targets for Federal sharpshooters and artillery.

About this time, General Gregg received a ball through the neck and was instantly killed. Col. Frederick Bass of the 1st Texas assumed command of the Texas Brigade until he himself was wounded. Command of the brigade then devolved upon Lt. Col. C. M. Winkler, the commander of the 4th Texas. Finding the brigade unsupported, Winkler led the remains of the command away from the murderous fire to the cover of woods and hollows in their rear. At noon, the Confederates finally broke off the engagement and retired to the west bank of Cornelius Creek, about five miles southeast of Richmond. The disastrous Battle of Darbytown Road cost the Texas Brigade 119 casualties (11 killed, 90 wounded, and 18 missing) out of the approximately 450 men engaged. The 5th Texas appears to have fared the worst, as only a handful of officers and 61 men were afterwards present for duty. Company F of the 5th Texas lost 10 of the 14 men who went into the battle. After the debacle, the remnant of the Texas Brigade took position on the Richmond defense line about four miles southeast of the city between the Williamsburg and Charles City Roads.

November 1864 - March 1865

After the engagements of October, the remnants of the Texas Brigade began settling into winter quarters along the Richmond defenses between the Williamsburg and Charles City Roads. As they had done in previous winters, the Texans built a shantytown of huts, tents, and log cabins. The only building material available was that derived from deserted homes in the area.

Simpson notes that "the Confederate cause appeared hopelessly lost by mid-winter of 1864-65. Sherman had reached Savannah, cutting the Confederacy in two; Grant had tightened the noose around Petersburg; Jubal Early had been driven from the Shenandoah Valley; Fort

Fisher, the last Confederate stronghold on the southeastern coast, had fallen; and the Union blockade was chocking the South to death."

April 1865

On April 1 the Confederate position at Petersburg completely collapsed. Grant ordered a major offensive all along his line on April 2. General Lee, seeing no other choice, ordered his soldiers to evacuate the Richmond Petersburg lines. His forces retreated toward the valley of Appomattox. On April 1, the Texas Brigade was sent from the trenches near Richmond to Petersburg. They arrived by train on April 2 and immediately took the position of honor, protecting the rear of the Confederate army as they moved toward Appomattox. On April 4, the Brigade followed Lee's army to Amelia Court House on the Richmond and Danville Railroad. The move westward continued until the 9th of April but it became clear that Lee was flanked on all four sides and did not have the forces to mount a breakthrough. Word circulated on the 9th that Lee had asked for a meeting with Grant.[3]

On the morning of April 10 the Texans heard their officers read Lee's Farewell Address. The formal surrender took place on April 12, 1885. There were 4,700 original members of the Texas Brigade. Only 617 were left to receive pardons on the 12th. Hamby, a member of John Tatum's old company, researched his old regiment after the war's end. He concluded that the 4th Texas lost 316 killed in battle, 431 wounded once, 98 wounded twice, 19 wounded three or more times, and 143 died of disease giving a total of 1,007 casualties and a rate of 80.4 percent.[4]

---

3. The information in this summary was taken from a non-documented posting on the internet (no source, no author). Additional information has been added and other parts deleted. In some cases information was corrected. In order to simplify the presentation of the summary, footnotes were not inserted. Information already cited in earlier chapters (Polley, Giles, Davis, Hood and Simpson) were used, to include *The Confederate Capital and Hood's Texas Brigade* by Mrs. A. V. Winkler.

4. Harold B. Simpson, *Hood's Texas Brigade: Lee's Grenadier Guard* (Dallas: Alcor Publishing Company, 1983), p. 468.

Epilogue

Like so many other soldiers—wounded, suffering from debilitating illness and far from home and family—it is likely that the chaplain was John Tatum's only visitor during his final long weeks in the hospital.

Chaplain Davis's reflections provide a fitting end to John's story and evoke the feelings of so many families who would never again hear the familiar footfall of their loved one returning home.

> . . . Many, both officers and men, were absent from camp. Some of them were in the hospitals, while others were left to sleep on the battlefield in the soldier's grave—they will no more attend the roll-call of their companions, command the men on parade, nor march to the music of the fife and drum—nor shall we any more meet them in the private walks of life even when this cruel war is over. They will not return with us when we take up our line of march for our homes in the far west. No, they will not go—they have already got their discharge, sealed in blood upon the altar of their country, but they have not gone to the flowery Prairies of Texas. And though friends may often look for them, and listen for their foot-fall upon the threshold, it will not be heard—sad though.—But when we return, we will tell his father that he fell with his battle harness on—sword in hand, and his face to the foe, and died with "forward" on his lips. We will tell his sister that even in death his face was lit up with a living luster, which had burned there since the day that Butler's order of New Orleans was first read on dress parade—when he swore his strong right arm should avenge a sister's wrongs. Yes, we will tell his mother where he fell, and where we buried him. We will tell her that we wrapped his blanket around him, covered him with his martial cloak, and buried him in a soldier's grave. And to that loved one whose image he wore, we will return this ring, and tell her he was a gallant boy.[5]

[5] Davis, *op. cit.*, pp. 101-102.

SELECTED BIBLIOGRAPHY

Bennett, William W. *A Narrative of The Great Revival Which Prevailed in the Southern Army*. Harrisonburg, Virginia: Sprinkle Publications, 1878.

Burton, Brian K. *Extraordinary Circumstances – The Seven Days Battle*. Indianapolis: Indiana University Press, 2001.

Catton, Bruce. *This Hallowed Ground*. New York: Doubleday & Co., 1956.

Connelly, Thomas L. and Barbara L. Bellows. *God and General Longstreet*. Baton Rouge: Louisiana State University Press, n.d.

Davis, Chaplain Nicholas A. *Chaplain Davis and Hood's Texas Brigade*. Edited by Donald E. Everett. San Antonio: Principia Press of Trinity University, 1962.

Drury, Ian and Tony Gibbons. *The Civil War Military Machine: Weapons and Tactics of the Union and Confederate Armed Forces*. New York: Smithmark Publishers inc., 1993.

Dyer, John P. *The Gallant Hood*. Indianapolis: Bobbs-Merrill Co., 1950.

Foote, Shelby. *The Civil War, A Narrative, Fort Sumter to Perryville*. New York: Random House, 1958.

Freeman, Douglas Southall. *Lee's Lieutenants*. 3 vols. New York: Charles Scribner's Sons, 1944.

Hennessy, John J. *Return to Bull Run: The Campaign and Battle of Second Manassas*. New York: Simon & Schuster, 1993.

Hood, John Bell. *Advance and Retreat*. New Orleans: P. G. T. Beauregard, 1880.

Jones, Rev. J. William. *Christ in the Camp, or Religion in the Confederate Army*. Harrisonburg, Virginia: Sprinkle Publications, 1887.

Lasswell, Mary, ed. *Rags and Hope: The Memoirs of Val C. Giles, Four Years with Hood's Brigade, Fourth Texas Infantry, 1861-1865.* New York: Coward-McCann, Inc., 1961.

Longstreet, James. *From Manassas to Appomattox – Memoirs of the Civil War in America.* Bloomington: Indiana University Press, 1960.

Luvaas, Jay and Harold W. Nelson. *The US Army War College Guide to the Battle of Antietam – The Maryland Campaign of 1862.* Carlisle, Pennsylvania: South Mountain Press, Inc., 1987.

McDonough, James Lee. *Chattanooga – A Death Grip on the Confederacy.* Knoxville: The University of Tennessee Press, 1984.

Polley, J. B. *A Soldier's Letters to Charming Nellie.* New York: Neale Publishing Co., 1908.

________, *Hood's Texas Brigade.* New York: Neale Publishing Co., 1910.

Sears, Stephen W. *Landscapes Turned Red: The Battle of Antietam.* New York: Warner Books, Inc., 1983.

Simpson, Harold B. *Hood's Texas Brigade: Lee's Grenadier Guard.* Dallas, Texas: Alcor Publishing Company, 1983.

________, *Gaines Mill to Appomattox: Waco and McLennan County in Hood's Texas Brigade.* Waco: Texian Press, 1963.

________, *Texas in the War, 1861-1865.* Hillsboro, Texas: The Hill Junior College Press, 1965.

Stackpole, Edward J. *From Cedar Mountain to Antietam.* Harrisburg, Pennsylvania: The Stackpole Company, 1959.

Tucker, Glenn. *Chickamauga: Bloody Battle in the West.* Dayton, Ohio: Morningside Bookshop, 1984.

Winkler, Mrs. A. V. *The Confederate Capital and Hood's Texas Brigade.* Austin: Von Boeckmann, 1894.

Woodward, C. Vann and Elisabeth Muhlenfeld . *The Private Mary Chesnut – The Unpublished Civil War Diaries.* New York: Oxford University Press, 1984.

John Tatum's

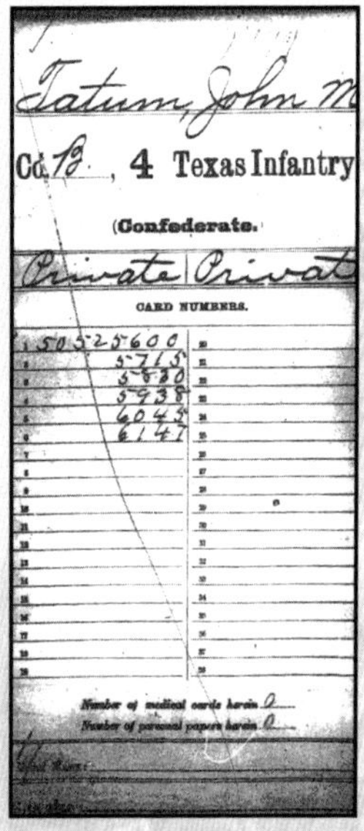
T. | 1119

Tatum, John M

Co. B, 4 Texas Infantry

(Confederate.)

Private | Private

CARD NUMBERS.

1 50525600
2 5715
3 5830
4 5938
5 6045
6 6147

Number of medical cards herein 0
Number of personal papers herein 0

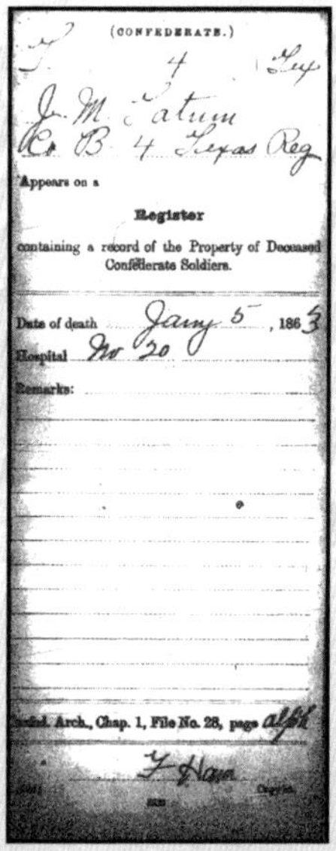
(CONFEDERATE.)

T. | 4 | Tex

J. M. Tatum
Co. B. 4 Texas Reg

Appears on a

Register

containing a record of the Property of Deceased Confederate Soldiers.

Date of death Jany 5, 1863
Hospital No 20
Remarks:

Confed. Arch., Chap. 1, File No. 28, page alph

F. Ham
Copyist.

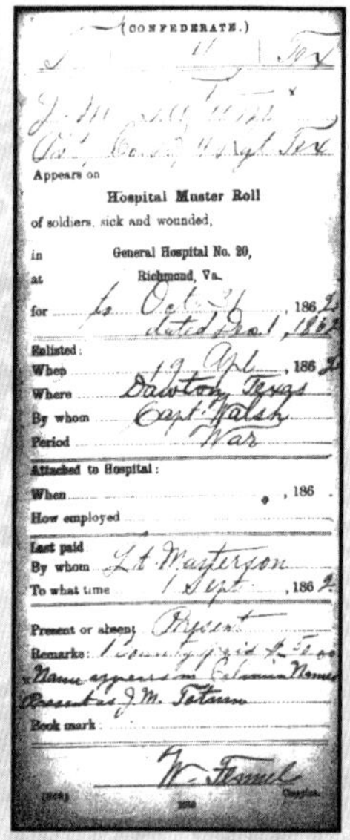
(CONFEDERATE.)

T. | 4 | Tex

J. M. Tatum
Pvt. Co. B. 4 Regt Tex

Appears on

Hospital Muster Roll

of soldiers, sick and wounded,

in General Hospital No. 20,
at Richmond, Va.
for Oct. 31, 1862
dated Dec 1, 1862

Enlisted:
When 19 Apl, 1862
Where Dawson Texas
By whom Capt Walsh
Period War

Attached to Hospital:
When , 186
How employed

Last paid
By whom Lt Masterson
To what time 1 Sept, 1862

Present or absent Present
Remarks: Bounty paid $50.00
Name appears in Column Names Present as J. M. Tatum

Book mark:

W. Fennel
Copyist.

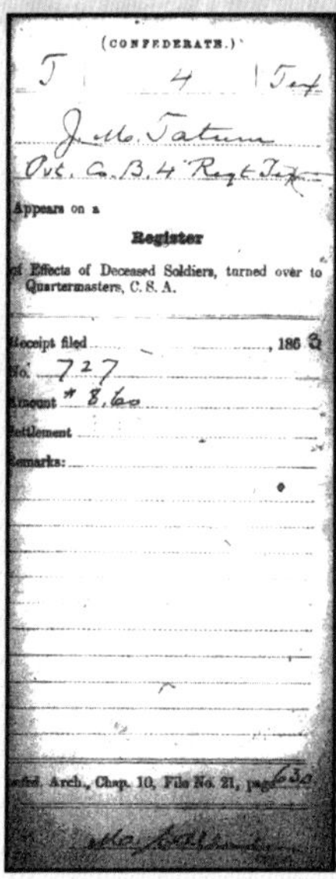
(CONFEDERATE.)

T. | 4 | Tex

J. M. Tatum
Pvt. Co. B. 4 Regt Tex

Appears on a

Register

of Effects of Deceased Soldiers, turned over to Quartermasters, C. S. A.

Receipt filed , 1863
No. 727
Amount $8.60
Settlement
Remarks:

Confed. Arch., Chap. 10, File No. 21, page 630

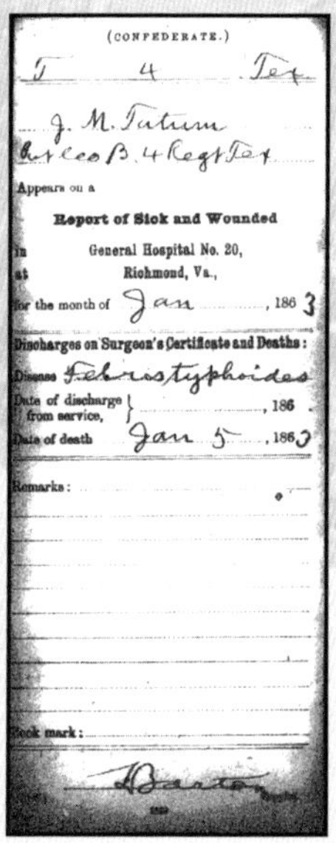
(CONFEDERATE.)

T. | 4 | Tex

J. M. Tatum
Pvt Co B. 4 Regt Tex

Appears on a

Report of Sick and Wounded

in General Hospital No. 20,
at Richmond, Va.,
for the month of Jan, 1863

Discharges on Surgeon's Certificate and Deaths:
Disease Febris typhoides
Date of discharge from service, , 186
Date of death Jan 5, 1863

Remarks:

Book mark:

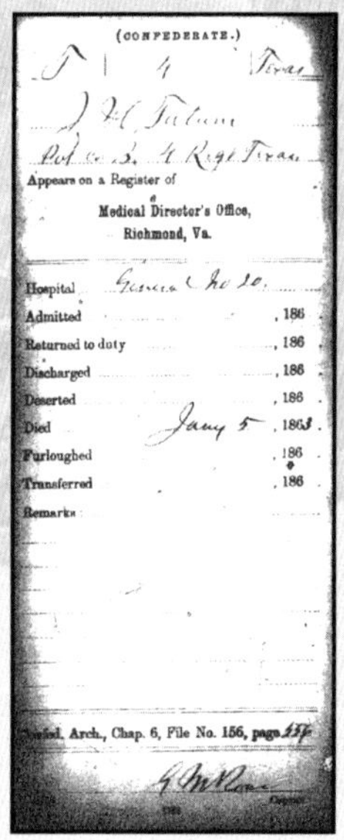
(CONFEDERATE.)

T. | 4 | Texas

J. M. Tatum
Pvt Co. B. 4 Regt Texas

Appears on a Register of
Medical Director's Office,
Richmond, Va.

Hospital General No 20
Admitted , 186
Returned to duty , 186
Discharged , 186
Deserted , 186
Died Jany 5, 1863
Furloughed , 186
Transferred , 186
Remarks

Confed. Arch., Chap. 6, File No. 156, page 457

Copyist.

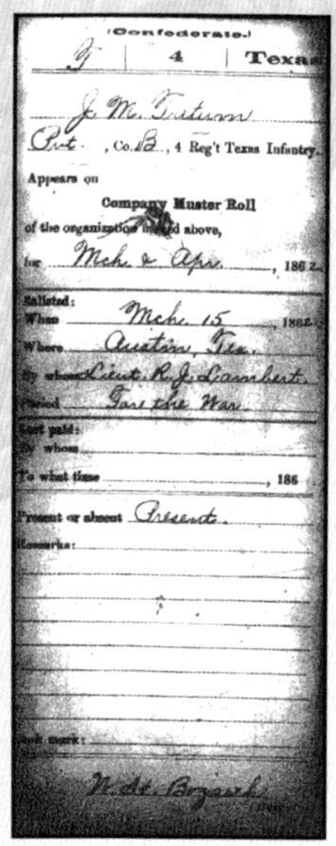
(Confederate.)

T. | 4 | Texas

J. M. Tatum
Pvt., Co. B, 4 Reg't Texas Infantry.

Appears on

Company Muster Roll

of the organization named above,
for Mch. & Apr., 1862.

Enlisted:
When Mch. 15, 1862.
Where Austin, Tex.
By whom Lieut. R. J. Lambert
Period For the War

Last paid:
By whom
To what time , 186

Present or absent Present.
Remarks:

Book mark:

W. H. Bozell

Military Record

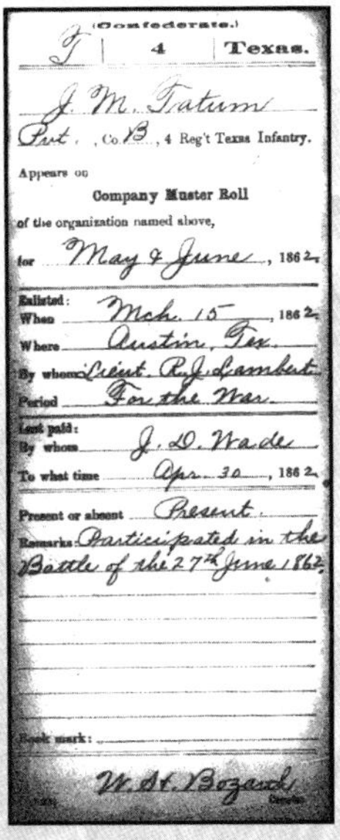
(Confederate.)
T | 4 | Texas.
J. M. Tatum
Pvt., Co. B, 4 Reg't Texas Infantry.
Appears on
Company Muster Roll
of the organization named above,
for May & June, 1862.
Enlisted:
When Mch. 15, 1862.
Where Austin, Tex.
By whom Lieut. R. J. Lambert
Period For the War.
Last paid:
By whom J. D. Wade
To what time Apr. 30, 1862.
Present or absent Present.
Remarks: Participated in the Battle of the 27th June 1862.
Book mark:
W. H. Bozarth, Copyist.

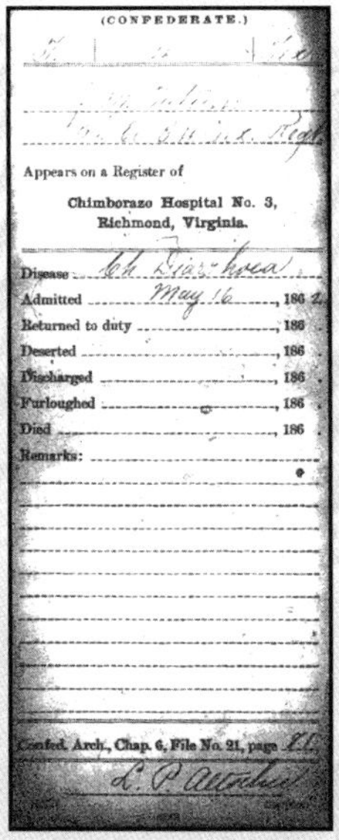
(CONFEDERATE.)
Appears on a Register of
Chimborazo Hospital No. 3,
Richmond, Virginia.
Disease Ch. Diarrhoea
Admitted May 16, 1862.
Returned to duty ..., 186
Deserted ..., 186
Discharged ..., 186
Furloughed ..., 186
Died ..., 186
Remarks:
Confed. Arch., Chap. 6, File No. 21, page 86
L. P. Attcher

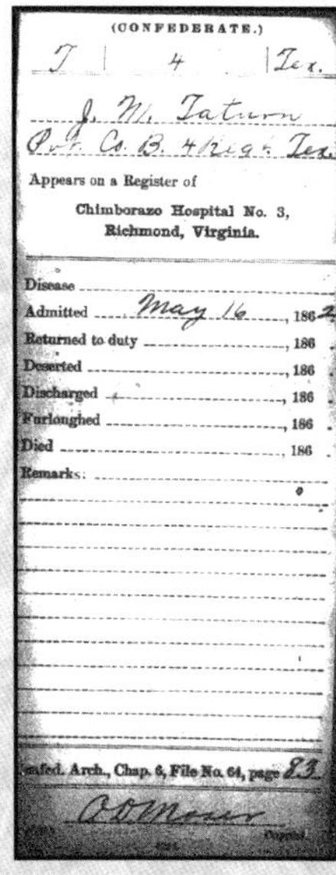
(CONFEDERATE.)
T | 4 | Tex.
J. M. Tatum
Pvt. Co. B. 4 Reg't Tex.
Appears on a Register of
Chimborazo Hospital No. 3,
Richmond, Virginia.
Disease
Admitted May 16, 1862.
Returned to duty ..., 186
Deserted ..., 186
Discharged ..., 186
Furloughed ..., 186
Died ..., 186
Remarks:
Confed. Arch., Chap. 6, File No. 64, page 83
A. D. Moore

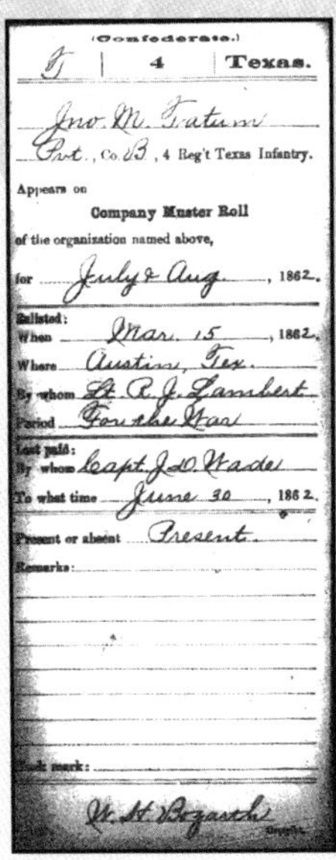
(Confederate.)
T | 4 | Texas.
Jno. M. Tatum
Pvt., Co. B, 4 Reg't Texas Infantry.
Appears on
Company Muster Roll
of the organization named above,
for July & Aug., 1862.
Enlisted:
When Mar. 15, 1862.
Where Austin, Tex.
By whom Lt. R. J. Lambert
Period For the War
Last paid:
By whom Capt. J. D. Wade
To what time June 30, 1862.
Present or absent Present.
Remarks:
Book mark:
W. H. Bozarth, Copyist.

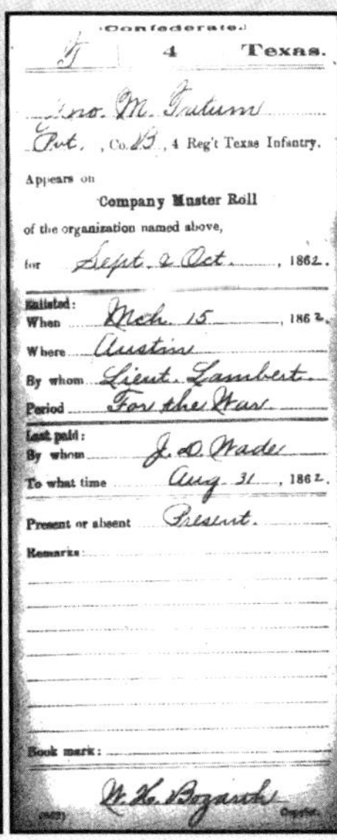
(Confederate.)
T | 4 | Texas.
Jno. M. Tatum
Pvt., Co. B, 4 Reg't Texas Infantry.
Appears on
Company Muster Roll
of the organization named above,
for Sept. & Oct., 1862.
Enlisted:
When Mch. 15, 1862.
Where Austin
By whom Lieut. Lambert.
Period For the War.
Last paid:
By whom J. D. Wade
To what time Aug. 31, 1862.
Present or absent Present.
Remarks:
Book mark:
W. H. Bozarth, Copyist.

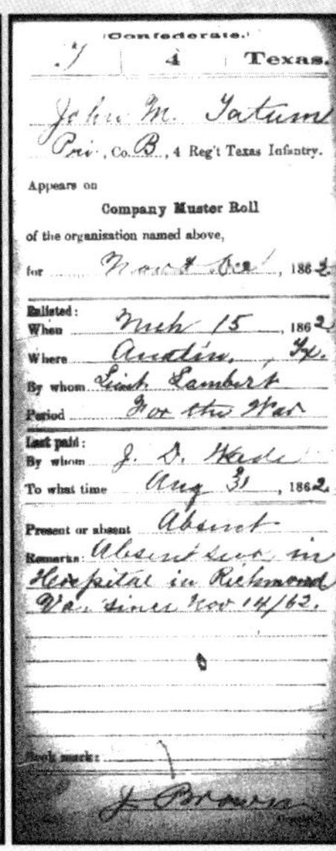
(Confederate.)
T | 4 | Texas.
John M. Tatum
Pri., Co. B, 4 Reg't Texas Infantry.
Appears on
Company Muster Roll
of the organization named above,
for Nov. & Dec., 1862.
Enlisted:
When Mch 15, 1862.
Where Austin, Tex.
By whom Lieut. Lambert
Period For the War
Last paid:
By whom J. D. Wade
To what time Aug. 31, 1862.
Present or absent Absent.
Remarks: Absent sick in Hospital in Richmond Va. since Nov. 14/62.
Book mark:
J. Brown

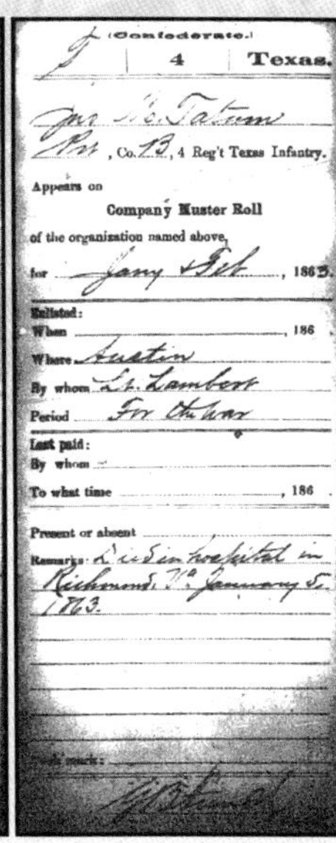
(Confederate.)
T | 4 | Texas.
Jno. M. Tatum
Pvt., Co. B, 4 Reg't Texas Infantry.
Appears on
Company Muster Roll
of the organization named above,
for Jany & Feb., 1863.
Enlisted:
When ..., 186
Where Austin
By whom Lt. Lambert
Period For the War
Last paid:
By whom
To what time ..., 186
Present or absent
Remarks: Died in hospital in Richmond, Va. January 5, 1863.
Book mark:

Appendix B
Tatum Family Narrative

John M. Tatum was born about 1827 in North Carolina (based on 1860 census). Tatum's were early arrivals in North Carolina, arriving from Virginia, especially the area around Jamestown. Early records show a Tatum family arriving in the Jamestown area from England in the very early years of America.

For reasons unknown at this time, John traveled to Pennsylvania and married Margaret Emerick Slack. According to the 1850 census, Margaret was married to Henry Slack, a blacksmith living in Pennsylvania about thirty miles south of State College. They had two children, John and Mary. In 1850, Henry was 29, Margaret 25, John 2, and Mary 7 months. Living with them at the time was Barbara Emerick (22), most likely Margaret's younger sister. According to "Slim" Tatum, Margaret's maiden name was Emerick and she was from Pennsylvania. Slim didn't know her first name or that she was married to someone else before she married John. Besides Margaret's younger sister, the census shows two others living in the house at the time: Samuel Alexander (4), and John Cotti (20) also a blacksmith. If these two were related to Henry or Margaret in some way is unknown.

Henry must have died about 1852 or 1853. There is a slight possibility that they divorced but divorce was very rare in those days and death from various ailments was common. John Tatum married Margaret in Pennsylvania somewhere between 1855 and 1857.

"Slim" Tatum remembers his dad telling him that Slim's grandfather and grandmother (John and Margaret) were married in Meadville, Pennsylvania. State and county records were not kept in those days. Any record would have been recorded by a local church and is yet to be discovered.

John moved Margaret and her two children to Pilot Point, Denton County, Texas, (just north of Dallas) where he worked as a blacksmith. It is possible that John learned his blacksmith trade from Henry Slack, something we will never know for sure. By 1860, John and Margaret added two children to the marriage: Margaret (named after her mother) and James. According to the 1860 census, John Tatum was 33, Margaret was 33 (although this number is probably off), John Slack was 12, Mary Slack was 10, Margaret Tatum was 2, and James B. Tatum was 2 months old. There is no record that John owned slaves (often this was noted on the census record).

The 1860 census reveals that several McBee families were neighbors of John and Margaret. "Slim" and "Kingfish" both indicated that John's friend prior to the Civil War was William McBee, also known as "Will." The census (Image 35, 1860 Census, Denton County, Pilot Point Post Office, Household 235) states that William McBee was born about 1834 in Tennessee. His wife, Louisa, was born the same year in Alabama. Although the couple would go on to have more children, two are listed: Francis M., born in 1849 in Marion, Tennessee; and, Rebecca J., born about 1853 in Missouri. "Slim" Tatum, based on information from his father, thought that Will or William fought with John Tatum during the Civil War. However, there is no indication that Will McBee was ever a part of the 1st, 4th, or 5th Texas Infantry Regiments (Hood's Brigade). It is possible that he was a part of another Texas unit. Will's son, Francis Marion, became James Tatum's father-in-law. "King-

William "Will" McBee

fish" remembers "Great-grandpa Will" as a large man with a beard that ended at his belt. He always carried a pocket watch on a long chain. Will died on December 23, 1918, as a result of a three month battle with Nephritis. At the time of his death he was one month short of 90. He is buried in Del Rio, Texas.

Living next door to William was James N. McBee, the brother of Will McBee, and his wife Martha (Household 236). James and Martha were both born the same year in Tennessee (1836). At the time of the 1860 census they had five children all born in Arkansas:

Marion Francis McBee,
Son of "Will" McBee and Louisa Price

William, 1851

Elijah, 1853

Letty A., 1855

Jasper, 1856

John J., 1858.

Rebecca McBee lived next door to William and James (Household 234). The 1860 census indicates that she was 78, born in 1782 in Tennessee. Her husband, also William McBee, had passed away before the 1860 census. Living with her at the time were Jasper McBee, 22 (born about 1838) in Tennessee, and Mary Jane McBee, 17 (born about 1843) in Missouri. Rebecca was the mother of William, James, and Jasper. Was the Jasper living with Rebecca William's brother? If so, Rebecca would have given birth at an advanced age. There is no indication as to how Mary Jane was related to Rebecca, perhaps a grandchild.

There was another McBee family living in the next county in 1860 (Grayson County, Texas), not far from Pilot Point. Jesse McBee, 48 (born about 1812), was married to Rebecca, 47 (born about 1813).

Jesse was born in Tennessee, Rebecca in Kentucky. All of their children were born in Missouri:

Eliz- (Female, probably short for Elizabeth), 1844

Jas A. (Probably "James"), 1848

Sami, 1850

Caroline, 1854

Tos J. (Male, may be short for "Thomas"), 1858.

Francis Marion McBee, son of William McBee, later married America Jeffers. Francis, better known as Marion, died on October 11, 1922, and is buried in Del Rio, Texas. All of their children were born in Texas:

Salina McBee (James' future wife), 1869

William R. McBee, 1875

Olive J. McBee, 1877

Samuel M. McBee, 1882

Laura A. McBee, 1885

David J. McBee, 1887

Charles McBee, 1897.

All of the McBee boys were large men; David was 6 foot, 6 inches. Their father, Marion, was not only tall but robust (about 275 pounds). Based on where and when the children were born, some of the McBee's went from Tennessee, to Missouri or Arkansas, and then to Texas. The evidence indicates that they didn't arrive in the Denton County area until 1859. It is more difficult to pinpoint John Tatum's arrival in the county. The most probable dates range from 1857 to 1858. Did John know the McBee's prior to Texas? The answer may be found with further research. Since John was born in North Carolina and most of the McBee's his age were born in Tennessee, the answer is not obvious. Many of the McBee's moved to Atascosa and Frio counties, southwest of San Antonio, sometime between 1863 and 1869.

Texas seceded from the Union on March 4, 1861, and immediately started raising several regiments for battle in Virginia. John did not initially volunteer to join the southern cause due to the death of his wife. Margaret was pregnant with a fifth child in 1861 and later died in childbirth. We assume that both the mother and child died as a result of the ordeal. When Lieutenant Lambert, Company B, 4th Texas Regiment, returned to Texas in early 1862 to recruit replacements, John signed on to serve (the Texas regiments lost a large number of men during the winter of 1861-62 due to illness). When he departed, he left his children with the McBee clan.

John fought as an infantryman with B Company in three major battles: Gaines Mill, Second Manassas, and Antietam (also known as Sharpsburg). He enlisted in March or April of 1862 and died of Typhoid in Richmond Hospital #20 (also known as the Alabama hospital) on January 4, 1863. He is buried in an unmarked grave in Oakwood Cemetery on the east side of Richmond.

What happened to the children after John's death? It is a difficult question to answer except for James. There are some assumptions that seem to make sense. John and Mary Slack were probably cared for initially by a McBee family. After John's death they may have been sent back to family members in Pennsylvania. Census records indicate that they had Slack as their last name when the Civil War started. Margaret is somewhat of a mystery. "Slim" and "Kingfish" Tatum remembered that their father, James, occasionally talked about a sister but they had no information concerning her whereabouts after the war. Early in the twentieth century, James took out several advertisements in an effort to find his sister. Unfortunately he didn't receive any response. If Margaret was raised by a McBee, it was not the same family that raised James. Whoever took her in probably adopted her and changed her last name making it difficult, if not impossible, to trace her whereabouts after 1860.

"Kingfish" Tatum recalls as a young boy meeting Aunt "Sophronzie," the lady that raised his father James. He vividly remembers her sitting in a rocking chair near the fireplace smoking a clay pipe.

She was living at the time somewhere between Del Rio and San Antonio, Texas, (most likely Frio County). "Sophronzie" is a nick name for Sophronia. The 1870 Census shows a Jesse and Sophronia McBee living in Atascosa County, Texas (south of San Antonio). According to this census, Jesse McBee was 53 and born in Alabama (other census records say Tennessee); Sophronia was 40 and born in Missouri. They had three children living in the home: Nancy F., 15, born in Missouri; James, 10, born in 1860 in Texas; and Jennie McBee, 2, born about 1868 in Texas. This is the only "Sophronia" mentioned in the 1870 census living in Texas. We know that James, the son of John and Margaret Tatum, was born in Texas in 1860, that he was raised by a Sophronia McBee, and would have been 10 when the 1870 census was taken. This has to be the "Aunt Sophronzie" that raised James Tatum and the "James McBee" mentioned in the census has to be James Tatum. Jesse and Sophronia were still living in Atascosa County according to the 1880 census, but the twenty-year-old James had moved on.

The 1880 census for Kinney County Texas (Kinney is a short distance east of Del Rio) shows a James Tatum, in his twenties, living with a L. F. Bell, his wife, M. J. Bell, and their three children. James is listed as a boarder and a laborer. The next door neighbor to the Bell's is W. A. McBee (probably Will McBee's nephew), 28 years old, and his wife Texana, 27 years old. Four children are listed ranging from 9 years to 2 months, all sons (J. A., Jessee, B. T., and Marcus). It is reasonable to assume that this is part of the McBee clan that had been a part of James' life.

It is interesting to note that in 1910 Sophronia was living with her daughter, Jennie McBee Bush, and son-in-law, J. T. Bush, in Precinct 2, Frio County, Texas. Frio is one county west of Atascosa County and a few miles closer to Del Rio. Obviously, her husband, Jesse, had passed away. The census shows her age as 69. This would have been about the same time that "Kingfish" remembers meeting Aunt Sophronzie somewhere near Del Rio. Of the four oldest children belonging to James Tatum, three of them were born in Frio County. This is further evidence that James was raised by Sophronia and never lived far away from her in the early days of his marriage.

James Tatum
Son of John M. and Margaret Tatum

Louisa McBee Tatum
James' wife and daughter of Marion McBee and America Jeffers

James Tatum married Louisa Salina McBee in 1886. Louisa was the granddaughter of William and Louisa Price McBee (neighbors and friends of John Tatum in Pilot Point), and the daughter of Francis Marion McBee and America Jeffers McBee. The 1850 census shows that William and Louisa were living in Marion County Tennessee; the 1860 census indicates that they were living in or near Pilot Point, Denton County Texas; and the 1870 census locates them in Atascosa County, Texas, Precinct 2, Somerset Post Office.

James and Louisa Salina McBee Tatum raised twelve children:

Ilene America Tatum, Born Dec. 15, 1886, in Pearsall, Texas

Marion David "Kingfish" Tatum, born June 29, 1889, in Rock Springs, Texas

Margaret Buella Tatum, born in 1891, in Pearsall, Texas

Julia Edna Tatum, born February 26, 1894, in Goldfield, Arizona

Rupert Franklin Tatum, born April 20, 1896, in Pearsall, Texas

Harry Charles Tatum, born January 14, 1898, in Rock Springs, Texas

Leonard Laverne "Slim" Tatum, born March 1, 1900, in Rock Springs, Texas

Bertha Elaine Tatum, born September 22, 1901, in Juno, Texas

Claude Hudspeth Tatum, born December, 1904, in Juno, Texas

Clyde Robert Tatum, born September 19, 1909, in Grand Falls, Texas

Alvin James Tatum, born October 9, 1911, in Grand Falls, Texas.

Some have wondered how Julia Edna Tatum, known to most people as Aunt Edna, was born in Arizona rather than Texas. "Kingfish" says that his dad packed up the family and headed for Los Angeles, California, with a covered wagon, a team of horses, four good mules, and a gray mare. They only made it as far as Arizona when they ran out of money. James Tatum went to work clearing land with his four mules while also working for a surveying crew laying out streets in or near Phoenix. Some of the streets were named after members of the crew. James would be surprised to learn that his street is now a major thoroughfare in the Phoenix area (Tatum Boulevard). While in Arizona, Julia Edna was born in Goldfield, Arizona, a present day ghost town just east of Phoenix. A little later the family returned to Pearsall, Texas, where Rupert was born (the family moved from the Del Rio area to Grand Falls in 1909). During this Texas phase, James owned a saloon that was often visited by some of the roughest cowboys in Texas.

The second effort to reach California came in 1911 when Alvin was a few months old. "Kingfish" felt sure that he could find every place that they stopped for the night during their journey. Their mode of transportation had not changed—horses and wagons. They arrived in Calexico, California (part of the Imperial Valley in Southern California), after sixty days on the trail. James Tatum and his son Marion (Kingfish) are noted on an early record as registered voters in Calexico. Later the family moved to Palo Verde, California, a few miles south of Blythe where many of the children picked cotton. James died on September 23, 1921, in Blythe of a sudden heart attack. His wife, Louisa Salina McBee Tatum, lived with various children until her death on January 21, 1939, at Loma Linda Hospital (now a major medical center) in Loma Linda, California. She is buried in Perris, California, where many of her children and their spouses are buried.

Below is a brief summary of what happened to the twelve children:

1. Ilene America Tatum married Alfred Charles Kessler in 1910 in Del Rio, Texas, lived most of her life in the central valley of California, died on February 5, 1980, and is buried in Kern County, California. She was given her grandmother's first name as her middle name.
2. Marion David "Kingfish" Tatum married Elsie May Dunagan on July 28, 1909, in Arizona. They lived most of their life in Bishop, California, after moving there from Nuevo (near Perris, California). Kingfish spent the majority of his life farming and raising cattle. He died in Bishop on December 22, 1975, and is buried in Perris.
3. Margaret Buella Tatum died in 1918 during the tragic influenza epidemic that hit the United States. She is buried in Blythe. She was married to Fred Ivey.
4. Julia Edna Tatum married Aaron Brooks, and after his death she married his younger brother William Kirby Brooks. She died in 1983 at the age of 89. She is buried in Ontario, California.
5. Rupert Tatum's first wife, Dorothy Wilford, died while they were living in Blythe. He later married Margaret Peel and moved to Independence, California (south of Bishop) in 1930. In 1960 they moved to Victorville, California, where

Rupert worked with his brothers Clyde and Alvin in the construction business. Rupert died in November, 1983, and is buried in the military cemetery in Riverside, California. He was a veteran of World War I.

6. Harry Charles Tatum was a farmer most of his life. He was married to Buela Logan. Harry joined his brothers Clyde and Alvin in farming potatoes and various other crops in Neuvo and Hesperia, California. Most of their farming was on leased land in Hesperia. Later Harry and Buela retired to Arizona. He died on January 16, 1983, and was buried in Mesa, Arizona.

7. Leonard Laverne "Slim" Tatum lived the majority of his life in Bishop farming and raising cattle. He was married to "Hazel" Schober and had one son, Jack. Slim died in June, 1983, and is buried in Bishop. Jack still has the saddle that his father used when the family made the sixty day trip to California in 1911.

8. Bertha Elaine Tatum married Irwin W. Wolford. She died in 1952 and is buried in Bishop, although she lived most of her life near Delano, California.

9. Irene Katherine Tatum married Fred Stump. They married in Blythe where they both lived most of their adult lives. Irene died on May 23, 1989.

10. Claude Hudspeth Tatum never married. While in his twenties, as the story is told by family members, he drank a bad bottle of homemade whiskey. The damage to his brain was severe enough that he had to be placed in facilities for the mentally handicapped. He spent most of his life in San Bernardino and Riverside, California. He lived to the ripe old age of 96. He died in 2000 in Riverside and is buried in Perris.

11. Clyde Robert Tatum was born as Clyde Irwin but had his middle name legally changed later. He married Nina Loretta Golding in Yuma, Arizona, in 1934. He started his ca-

reer farming with his brothers in Nuevo, California. About 1946, Clyde along with Harry and Alvin moved their farming operation to Hesperia, California. Clyde started building homes in the fifties and continued building until his retirement. His first wife, Nina, died in June, 1960, and is buried in Perris. Clyde and his second wife, Bunny, are buried in Victorville. He died on January 21, 1996.

12. Alvin James Tatum was married to Julia Golding, sister to Nina. Alvin worked with Clyde in the farming operation in Hesperia, as well as the construction business in Victorville. He was admitted to Loma Linda Hospital in 1967 with a severe case of peritonitis and died a few weeks later. He is buried in Perris. He was only 55 years old at the time.

Recommended areas for further research:

1. Where was John Tatum born in North Carolina? Can we trace his ancestors back to the area around Jamestown? Census records prior to 1850 only listed the head of the household making it difficult to find John's father or grandfather.
2. Is there a record in Pennsylvania for John and Margaret Tatum's marriage? Marriage records in the 1850's were kept by the church.
3. What happened to John and Margaret's daughter Margaret? Was she raised by a McBee? A place to start is those neighbors living near John in 1860 (available in the 1860 census). Census records from the 1800's and early 1900's are full of errors. Birthplaces, spelling of names, ages, and other errors are common. For example, the 1900 census lists James Tatum as "James Tadem."

Tatum/McBee Family Tree

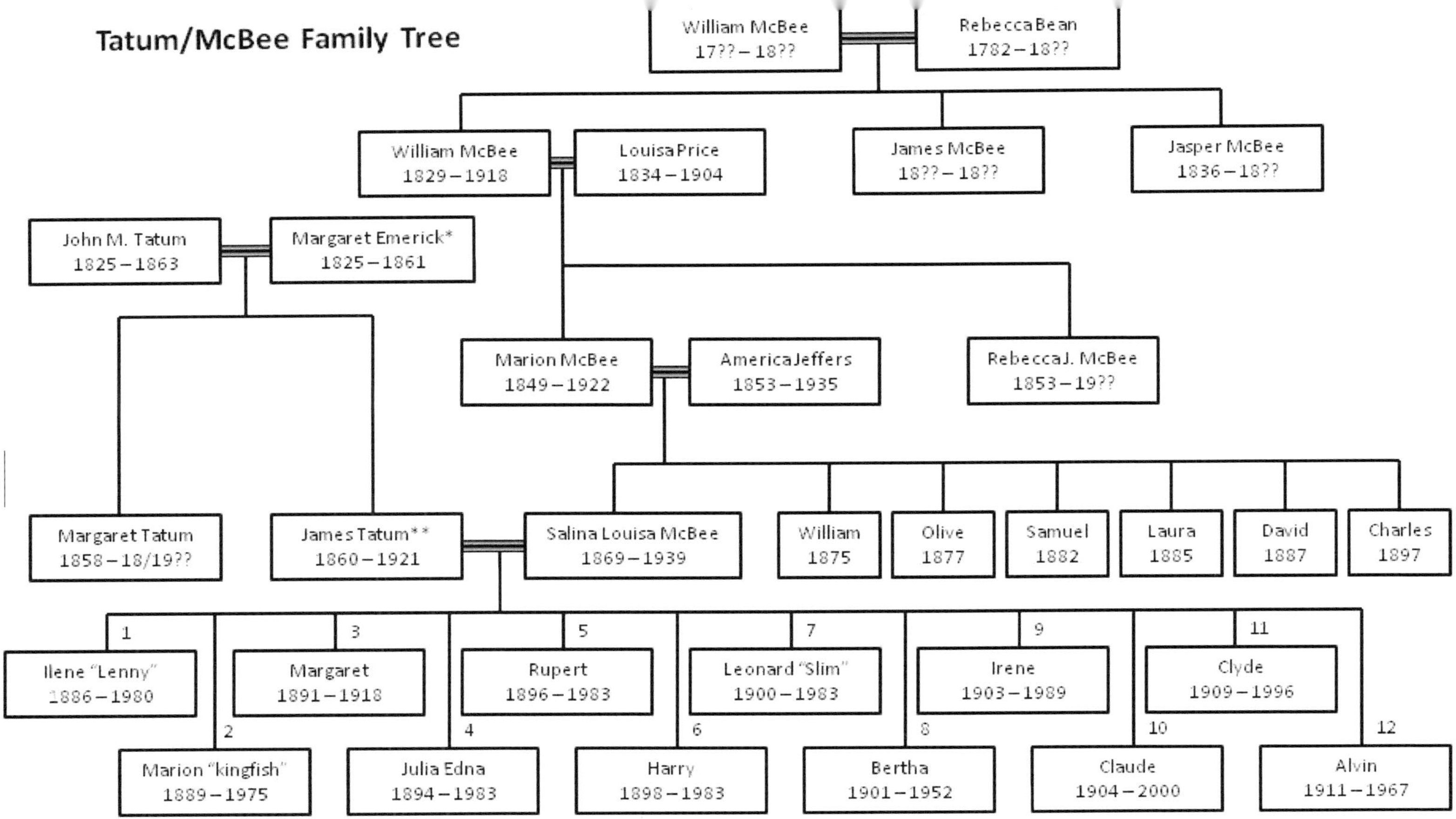

•Margaret Emerick Tatum was previously married to Henry slack. They had two children: John Slack – 1848, and Mary Slack – 1850.

**James Tatum was raised by Jesse and Sophronia McBee. His mother died in late 1861 or early 62 during childbirth; his father died in the Civil War. James never knew who raised his sister after his parents died and was never able to locate her.

James Tatum, son of John M. Tatum, during his early days in Texas

Appendix C

VISITING THE BATTLEFIELDS

What to bring

Bring your camera, laptop computer, a guide to the Civil War Battlefields (available at most book stores), casual clothes for hiking, and this book.

Day One

Fly into Dulles International, Reagan National, or Baltimore-Washington Airport (look for the cheapest flight—fortunately you have three choices).

Rent a car and drive to Richmond, Virginia, about a 90 minute drive from Washington, DC, south on Interstate 95.

Spend the night in Richmond (recommended hotel for a feel of old Richmond –Jefferson Hotel – not the cheapest place in town but a great experience. Numerous other motels/hotels are available outside of the city).

Day Two

Visit the Gaines Mill battlefield (take exit 34-Creighton Road-east off of Interstate 295, turn right on 156-Cold Harbor Road-to Gaines Mill

At the battlefield, park in the small parking lot in front of the only house standing. Take the path through the woods that starts at the parking lot. Try to imagine that most of the trees were not there during the attack. The path is in the middle of the Union lines. In fact, if you look carefully you will see where the Union soldiers dug small defensive positions.

Halfway down the path, you will see a path leading to the top of the hill. This is where the 4th Texas and John Tatum made their attack and breakthrough. If you take the path to the top of the hill you will find a marker to Whiting's Division (the 4th Texas was part of this division during the battle). Return to the main path and take it to the end. It will circle back to the beginning.

While you are in the vicinity, visit the Cold Harbor battlefield nearby. Cold Harbor was one of the last big battles of the Civil War. The 4th Texas fought here but many months after John's death.

Day Three

Visit Chimborazo Hospital where John spent some time shortly after arriving in Virginia. There is a small museum on the grounds. Try to imagine countless buildings on these grounds with patients running into the hundreds, if not thousands. Walk to the edge of the bluff overlooking the James River. Here is where the chamber pots were poured onto the open ground. You can imagine the smell as the prevailing breeze carried it into the wards.

Visit the site where John died, the former location of Hospital No. 20, also known as the Alabama Hospital. It is located at 25th and Franklin Streets. There are buildings still standing nearby that resemble the original hospital.

Drive north on 21st Street to Oakwood Cemetery where John is buried. You will find the unmarked graves of most of the Confederate soldiers on the far side of the cemetery.

Spend of the rest of the day visiting historical museums in Richmond (there are many).

Possible side trips

Colonial Williamsburg and James Town are an hour's drive east of Richmond on Interstate 64.

The battlefields near Petersburg are only a 30 minute drive south on Interstate 95. The place of surrender, Appomattox Court House, is west of Petersburg.

There are numerous other battlefields on the outskirts of Richmond.

Day Four

Get up early and drive to the battlefield in Manassas. Take Interstate 95 north to Highway 234. Highway 234 exits 95 just north of Quantico Marine Base. It is worth a stop at the new Marine Museum right next to 95 and about three miles south of the 234 exit.

Stop at the Manassas battlefield visitor's center, pay the fee, and get a guide to the battlefield. The first battle of the Civil War was fought on this battlefield. John Tatum and the 4th Texas were part of the second and much larger battle. Be sure to visit the site where the 5th Texas almost completely annihilated the 5th New York. The memorial marker to the 5th New York stands on the original site. While there, you can see some of the terrain across the creek (Young's Branch) where the 4th Texas charged the Federal guns belonging to Captain Kerns. There is a path that leads from this spot to Chinn Ridge (1 ½ miles) that duplicates the direction taken by the Texas Brigade. You can also reach Chinn Ridge by car. Be sure and visit the old stone bridge and the railroad cut where Jackson fought with great determination. Remember, the majority of the trees that you see were not there when the battle was fought.

Day Five

Visit Thoroughfare Gap, about 10 miles east of Gainesville on Highway 55. If you are visiting on a Saturday or Sunday, take Highway 55 west through Haymarket to right on Turner Road. Immediately after crossing over Interstate 66 turn left on Beverly Mill Road. At the end of the road you will find the old stacked stone mill built by slaves. It played an important part in the battle for the gap. The mill is not open on week days.

Visit old town Manassas and the Manassas Museum. If you like old Civil War towns, drive thirty miles northwest to Leesburg.

Possible side trips

Washington, DC, is 28 miles east of Manassas on Interstate 66. It is easy to spend anywhere from one to seven days seeing the museums, memorials, historic sites, etc. Park your car and use the subway system. It is easy to use.

Annapolis, Maryland, and the Naval Academy are two hours east of Manassas. The old capital is full of historic buildings.

Day Six

Take Highway 234 north from Manassas to Highway 15 north. Highway 15 will take you all the way to Frederick, Maryland. From 15, turn right onto Interstate 70 west. A few miles after getting on 70, take the exit for Highway 40 alt. westbound. Highway 40 is the old National Pike and will take you through South Mountain where the 4th Texas fought a delaying battle while Lee's troops moved across Antietam Creek near Sharpsburg. There are historical markers near the road showing where the battle was fought. Continue on 40 to Boonsboro then left on Highway 34 to the battlefield at Antietam.

Stop at the visitor's center, pay your fee, and pick up a guide to the battlefield. From the visitor's center, walk north to the Dunkard church. The 4th Texas spent part of their time resting in the woods behind the church. Walk further north to the "bloody" cornfield. On the west side of the cornfield and on both sides of the fence next to the road, the 4th fought one of their bloodiest conflicts. The 1st Texas fought in the middle of the field and the 5th Texas on the right side of the field. There is a marker to the Texas Brigade on the south end of the field.

Return to your car and follow the rest of the battle from your car. If you follow the battle's progression you will end up near Burnside Bridge. It is a good place to see some of the memorials above the bridge. One memorial was dedicated to President McKinley when he was a soldier in the union army, commending him for bringing coffee to the soldiers!

Spend the night in Harper's Ferry or Shepherdstown, West Virginia, or Hagerstown, Maryland. All three of them are a short drive. There are several B&B's in the area, including two on the battlefield.

Possible side trips

The Gettysburg battlefield is an hour's drive north. Take Interstate 81 north to Highway 30 west. It will take a full day to see the battlefield. See if you can find the marker where the Texas Brigade began their attack on the second day of the battle. It is the same design as the one at Antietam.

Visit the historic town of Harper's Ferry. It is a short drive south of Sharpsburg on Harper's Ferry Road.

Day Seven

Return to Washington, DC, for your flight home.

ABOUT THE AUTHOR

Timothy C. Tatum is the great-grandson of John M. Tatum. Tim served for 29 years as a chaplain in the United States Army. He holds a doctorate from the University of Southern California along with master's degrees from Talbot Theological Seminary, Long Island University and Southwestern Baptist Theological Seminary. He is also a graduate of the US Army Command and Staff College and the US Army War College where he also served as the Director of Ethics.

Lieutenant Colonel Winkler, who was the last commander of the 4th Texas Infantry Regiment and was with them when they surrendered at Appomattox.